ANALYTICAL
PHILOSOPHY OF
HISTORY

In historic events the rule forbidding us to eat of the fruit of the Tree of Knowledge is specially applicable. Only unconscious action bears fruit, and he who plays a part in an historic event never understands its significance. If he tries to realize it his efforts are fruitless. (Tolstoy, *War and Peace*, book XII, ch. 2.)

Pour que l'évenement le plus banal devienne une aventure, il faut et il suffit qu'on se mette à le *raconter*. C'est ce qui dupe les gens: un homme, c'est toujours un conteur d'histoires, il vit entouré de ses histoires et des histoires d'autrui, il voit tout ce qui lui arrive à travers elles et il cherche à vivre sa vie comme s'il la racontait . . . J'ai voulu que les moments de ma vie se suivent et s'ordonnent comme ceux d'une vie qu'on se rapelle. Autant vaudrait tenter d'attraper le temps par la queue. (Sartre, *La Nausée.*)

ANALYTICAL
PHILOSOPHY OF
HISTORY

BY

ARTHUR C. DANTO

Associate Professor of Philosophy
Columbia University

CAMBRIDGE
AT THE UNIVERSITY PRESS
1965

PUBLISHED BY

THE SYNDICS OF THE CAMBRIDGE UNIVERSITY PRESS

Bentley House, 200 Euston Road, London, N.W. 1

American Branch: 32 East 57th Street, New York, N.Y. 10022

West African Office: P.O. Box 33, Ibadan, Nigeria

Library of Congress Catalogue
Card Number: 65-11205

Printed in Great Britain by
Spottiswoode, Ballantyne and Co., Ltd.
London and Colchester

TO
SHIRLEY

PREFACE

It sometimes is said that the task of philosophy is not to think or talk about the world, but rather to analyse the ways in which the world is thought and talked of. But since we plainly have no access to the world apart from our ways of thinking and talking about it, we scarcely, even in restricting ourselves to thought and talk, can avoid saying things about the world. The philosophical analysis of our ways of thinking and talking about the world becomes, in the end, a general description of the world as we are obliged to conceive of it, given that we think and talk as we do. Analysis, in short, yields a descriptive metaphysic when systematically executed.

It is impossible to overestimate the extent to which our common ways of thinking about the world are historical. This is exhibited, if by nothing else, by the immense number of terms in our language, the correct application of which, even to contemporary objects, presupposes the historical mode of thought. Should there be a people somewhere in time who truly thought unhistorically, we would know this by the fact that communication with them would be marginal, vast regions of our language being untranslatable into theirs. To endeavour, ourselves, to think unhistorically would require at least a linguistic restraint, for we would be committed to get on with a fragment of our vocabulary and grammar. We should, indeed, have to restrict descriptions to just those predicates which pass muster under empiricist criteria of meaningfulness. Empiricists, who have found this limited vocabulary uniquely meaning-ful, have encountered problems in connection with history, which is only natural, given the criteria they impose. It is the glory of empiricism to be austere, and it is by meeting and solving the problems which it raises that one begins to get a dim sense for the contours of historical thought, and hence for the structure of history itself. This book is an analysis of historical thought and language, presented as a systematic network of arguments and clarifications, the conclusions of which compose a descriptive metaphysic of historical existence.

More than this it would not be advisable to say by way of prefatory remark, and I say this much only to explain somewhat the title of the book and the spirit in which it was written. And to suggest, finally, the odd view I have that philosophy has things to say in its own right, that 'analysis'—which I employ eclectically (in the sense given that term by the Bolognese painters)—is the way in which to say them, and that the distance from Cambridge to Saint-Germaine-des-Près is not so astronomically vast as it appears.

The bulk of this book was written during a sabbatical leave from Columbia University, 1961–62, and supported by a fellowship award from the American Council of Learned Societies. To both of these institutions I am grateful for their tangible encouragement and aid. Earlier studies were abetted by two summer grants from the Columbia Council for Research in the Social Sciences. Two sections of the book—chapters VIII and XII—first appeared as articles in *History and Theory* and *Filosofia* (Turin; 4th International Fascicle, November 1962) respectively, and I thank the editors of these journals for permission to use this material again.

There are three persons to whom I should like to acknowledge an intellectual debt. The first of these is Professor William Bossenbrook, whose courses in history at Wayne University awakened me, and a whole generation of students, to the world of intellect. His lectures were the most stimulating I ever audited, and I should have devoted my life to the study of history as a result of them were it not for the discovery that they were unique. The second is Professor Ernest Nagel, whose work in philosophy of science, and especially in the topic of reduction, has been paradigmatic of high philosophical achievement. I have profited from his example and his encouragement. The third is my close friend and colleague, Professor Sidney Morgenbesser, a man of warmth, wit, and extraordinary philosophical acuity. His own submission to the highest standards of philosophical integrity stands as a kind of conscience upon all who know him. My book bears the mark of all these three men.

Many others have stimulated my thought, sometimes in ways they might not recognize, and should they read this book they may find some phrase or thought of theirs cut out of a conversation and mounted like a fragment in a collage: I have not borrowed, but stolen, making their matter mine. I have, in addition, some special obligations to Justus Buchler, Robert Cumming, James Gutmann, Judith Jarvis Thomson, and John Herman Randall, to each of whom I shall always be grateful. My daughters, Elizabeth and Jane, furnished me with example after example of historical explanation, some of which are preserved in my discussion of that tormented subject.

To the extent to which this essay has any clarity or literary surface, I owe it to my wife, Shirley Danto, whose unerring eye and ear for literary rectitude and taste have been my guide. Where the writing is dark, this will be due to my having failed to consult or heed her. But of course my debt to her is, in every way, immense.

A.C.D.

NEW YORK
1964

CONTENTS

Preface   *page* 7

   I Substantive and Analytical Philosophy of History   I

  II A Minimal Characterization of History   17

 III Three Objections against the Possibility of Historical
      Knowledge   27

  IV Verification, Verifiability, and Tensed Sentences   34

   V Temporal Language and Temporal Scepticisms   63

  VI Evidence and Historical Relativism   88

 VII History and Chronicle   112

VIII Narrative Sentences   143

  IX Future—and Past—Contingencies   183

   X Historical Explanation: The Problem of General Laws   201

  XI Historical Explanation: The Rôle of Narratives   233

 XII Methodological Individualism and Methodological
      Socialism   257

Notes   285

Index   315

SUBSTANTIVE AND ANALYTICAL PHILOSOPHY OF HISTORY

Two distinct kinds of inquiry are covered by the expression 'philosophy of history'. I shall refer to these as *substantive* and *analytical* philosophy of history. The first of these is connected with ordinary historical inquiry, which is to say that substantive philosophers of history, like historians, are concerned to give accounts of what happened in the past, though they are concerned to do something *more* than just that. Analytical philosophy of history, on the other hand, is not merely connected with philosophy: it *is* philosophy, but philosophy applied to the special conceptual problems which arise out of the practice of history as well as out of substantive philosophy of history. Substantive philosophy of history is not really connected with philosophy at all, any more than history itself is. This book is an exercise in analytical philosophy of history.

The first thing I shall analyse is what substantive philosophy of history pretends to do in addition to giving an account of the past. One might say, roughly, that in contrast with even the most ambitious piece of ordinary historical writing, a philosophy of history seeks to give an account of the *whole* of history. There are, however, some initial difficulties in this characterization. Suppose we took together all the pieces of ordinary historical writing, and to these then added further pieces of historical writing which filled in all the gaps, so that we had, in the end, a complete and total description of everything that has ever happened. It might then be said that we had produced an account of the whole of history, and hence a philosophy of history. But in fact we would not have done this: we would at best have produced an account of the whole *past*. We must, accordingly, distinguish between the whole of history and the whole past. And one way of doing it might be as follows.

Typically, we think of historians as interested in studying, and in writing accounts of, particular past events in very great detail. I use the word 'event' with some looseness here, but the French Revolution would clearly be an instance of the kind of event which historians are interested in studying and accounting for. Now there must be innumerable events for whose occurrence we have scant evidence, and a great many others which we believe must have happened, but about which we know little more than that they must have happened. There are, in short, many gaps in our account of the past. But just suppose that these gaps were all filled in, so that we knew as much about every event which ever happened as we know about the French Revolution. Let us, indeed, suppose that we know everything that ever happened; that we have some Ideal Chronicle of the whole past. This would still not be the whole of history with which I have said that substantive philosophers of history are concerned. Such an ideally complete account of the whole past would at best furnish *data* for a substantive philosophy of the whole of history. The concept of data is correlative with the concept of theory, and the plain suggestion here is that substantive philosophy of history is an attempt to discover a kind of theory concerned with the, as yet unclarified, notion of the whole of history. I shall follow this suggestion, and identify two distinct kinds of such theories, *descriptive* and *explanatory*.

A descriptive theory in this context is one which seeks to show a pattern amongst the events which make up the whole past, and to project this pattern into the future, and so to make the claim that events in the future will either repeat or complete the pattern exhibited amongst events in the past. An explanatory theory is an attempt to account for this pattern in causal terms. I am insisting that an explanatory theory qualifies as a philosophy of history only insofar as it is connected with a descriptive theory. There are any number of causal theories which seek to account for historical events in the most general terms—explainable by reference to racial or climatic or economic factors. But these theories are at best contributions to the social sciences, and are not, as such, philosophies of history. Marxism is a philosophy of history, and indeed exhibits both theories: the descriptive and the explanatory. Seen from the point of view of the descriptive theory, the pattern is one of class conflict, where any given class generates its own antagonist out of the conditions of its own existence, and is overthrown by it: 'all history is

2

the history of class struggles': and the shape of history is dialectical. This pattern will continue as long as certain causal forces are operative, and the attempt to identify these causal forces with various economic factors constitutes the explanatory theory of marxism. Marx predicted that the pattern would terminate at some future time because the causal factors responsible for its continuing will become inoperative. What would happen after that Marx hesitated, save in some cautious utopian hints,[1] to say. But then, he felt, the term 'history' would no longer apply. History, as he understood it, would come to an end when class-conflicts came to an end, as they would when society came to be classless.[2] And he, Marx, was only offering a theory of *history*.[3] At all events, it should be clear that the expression 'the whole of history' covers more than does 'the whole past'. It covers, as well, the whole future or, if it is important to make this qualification, the whole *historical* future. I shall revert to this in a moment.

If we see the connection between history and philosophy of history in the manner I have suggested, we might be tempted to understand this connection as analogous to the connection between observational astronomy and theoretical astronomy. Thus, Tycho Brahe is celebrated for having made, over a long period of time, a series of celestial observations of unprecedented accuracy concerning the positions (amongst other things) of the known planets. Yet he himself failed to find a projectable pattern amongst these various positions. It was Kepler who succeeded in this, discovering, after some arduous work, that a planet's positions could be located on an ellipse with the sun as one focus. This would be like having what I have called a descriptive theory. It remained for Newton to explain why this particular pattern held; that is, to offer an explanatory theory. On occasion, philosophers of history have seen their task in terms exactly analogous to these. Kant, for example, writes in this manner:

Whatever metaphysical theory may be formed regarding the freedom of the will, it holds equally true that the manifestations of the will in human actions are determined, like all other external events, by universal natural laws. . . . In view of this natural principle of regulation, it may be hoped that when the play of freedom of the human will is examined on the great scale of universal history, a regular march may be discovered in its movements, and that, in this way, what appears to be tangled in the case of individuals, will be recognized

3

in the history of the whole species as a continually advancing, though slow, development of its original capacities and endowments.... We will accordingly see whether we can succeed in finding a clue to such a history; and in the event of doing so, we shall leave it to nature to bring forth the man who will compose it. Thus did she bring forth a Kepler who, in an unexpected way, reduced the eccentric paths of the planets to definite laws; and then she brought forth a Newton, who explained these laws by a universal natural cause.[1]

Were we to continue with this somewhat flattering comparison, substantive philosophy of history would stand in the same relationship to ordinary historical inquiry that theoretical science stands to scientific observation. There have been, and perhaps there still are, parts of science which have not passed beyond the mere making of observations, the collecting of specimens, and the like. Ordinary history might be just such a science. Substantive philosophy of history might then constitute a step towards bringing history to the next two levels (the keplerian and the newtonian levels respectively) of scientific understanding. Indeed, 'philosophy of history' would be the science of history, and its being known as 'philosophy' would be simply a quaint survival of the older use of the term in accordance with which physics was once called 'natural philosophy'. Kepler's laws, though based on the data gathered by Tycho, went beyond them, enabling astronomers not merely to organize into a coherent pattern all the positions of the planets observed by Tycho but to predict all their *future* positions and even those of planets unknown in Kepler's time. Newton's laws did not simply explain the facts known to Tycho and to Kepler, but (ideally) a great many facts unknown to them. Similarly, it might be urged, a truly successful *historical* theory would go beyond the data gathered by history, not only reducing them to a pattern, but predicting, and explaining, all the events of future history. It might be said, then, that this is the sense in which substantive philosophy of history is concerned with the *whole* of history: the whole past and the whole future: the whole of time. Historians, by contrast, are concerned only with the past, and with the future only when it becomes past. For all our *present* data come from the present and the past: we cannot *now* gather data from the future: and history is *just* a data-gathering enterprise.

Such an account is exceedingly generous to substantive philosophy of

4

history. But it is singularly ungenerous to history itself. Even supposing philosophies of history were attempts at something like scientific theories, one cannot but conclude, from any acquaintance with them, that they are quite crude attempts, so crude, indeed, that when contrasted with even so simple a descriptive theory as Kepler's, existing philosophies of history are unspeakably inept, with almost no power to predict. Explanatory philosophies of history, even those which have been most influential, are little better than programmes for theories which remain to be formulated, much less tested. On the other hand, if we think of ordinary historical accounts (and not even just the best of these), they seem to be highly developed instances in their own genre, satisfying criteria applicable to that genre, and throwing into relief the way in which philosophies of history fail miserably to satisfy the criteria for a scientific theory.

Moreover, the genre, whose criteria historical accounts appear to satisfy, does not, on the face of it, include such things as sequences of inscriptions reporting planetary positions on successive nights. It is very difficult to class such a work as, say, Gibbon's *Decline and Fall of the Roman Empire*, with Tycho Brahe's observatory notes, or with any set of reports of scientific observation. Or rather, there exists within history itself something like the sort of activity with which history as a whole is compared in the account we are considering. I have in mind the sort of thing which is done when historians use special techniques to authenticate documents and artifacts, or to date an event, or to decide whether Sir Walter Raleigh really was an atheist, or to identify an individual. Such activities might indeed be usefully regarded as observational, yielding single sentences, hopefully true, such as 'Sir Walter Raleigh was not an atheist'. But this is by no means all that historical activity consists in. There are also, within history itself, attempts to organize the known facts into coherent patterns, and such organizations of facts have, in a way, very nearly as much in common with scientific theories as have philosophies of history. Of course they do not in quite the same way admit of projection into the future. But they nevertheless do have some predictive power. A certain account of what happened in the past, based on evidence, might allow us to predict some further fact about what took place which we were heretofore ignorant of: and independent investigation might confirm this prediction. The fact that the predicted occurrence took place in the past must not blind us to the fact that it was a prediction, and, if you like, a prediction about what we, as historians, will subsequently find if we make an investigation. And this is very like predicting what we will see in the heavens if we make a certain observation. Thus, finding three elaborate roman-style tombs in different parts of Jugoslavia, and knowing the Roman habit of burying people by the sides of roads, might suggest that these tombs all lay on some main road: and subsequent investigation might bear this prediction out. The distinction between observation and theory has, then, at least an analogue within history. There may be vast differences between historical accounts and scientific theories. But no vaster, one feels, than the difference between philosophies of history and scientific theories.

Further, it is inaccurate and distorting to think of historical writing as consisting of nothing but data for future philosophies of history (Tycho wanted to find a descriptive theory to fit his observations, but it is certainly false to suppose that historians see their own 'observations' in that light). It does not follow that what historians do could *not* be seen in that light, but only that they do not see it in that way, any more than artists see themselves as providing data for art-historians, even if it happens to be true that what artists do is, in fact, the data with which art-historians work. However we might, in a different context, characterize historical work, the present account does not describe this work in accordance with the purposes and criteria of achievement which are those of practising historians. And to accept this account would involve a revolution in our concept of history as an intellectual discipline. If I happen to read an account of the Thirty Years War which stimulates me to think about historical explanation, it would be true that the historian who wrote it stimulated some philosophical reflection. But it was not his purpose in writing it to stimulate reflection of this sort. What we do have, of course, is some such situation as this. An individual historian works very hard to establish, say, a certain fact concerning the past. And then some other historian finds a use for this fact in writing an account of some part of the past. This may or may not be a satisfactory account in the eyes of his peers. But if it is unsatisfactory, another account may be written, and an account of just the same sort as the one it replaces, but satisfying exactly the same criteria as those by which the other one was deemed unsatisfactory. Accounts of this kind (and I shall have more to say about the criteria historical accounts must satisfy) are somehow consummatory, in the sense that any improvement of them is still a production which remains within history. These accounts, in other words, do not seem to be preliminary to any other, different *kind* of activity, but only, perhaps, to further accounts of just the same kind, satisfying just the same criteria.

The difference, then, between history and a philosophy of history cannot be that the latter gives, as the former does not, accounts based upon detailed factual findings. For such accounts are given by history and philosophy of history alike. So the account given by a philosopher of history would have to be a quite different kind of account if it is to get outside the area of history, and do something other than what history itself does. And of course one would expect it to be quite a different kind of account if it were at all like a scientific theory, for scientific theories seem, on the face of it, to belong to a different genre and to satisfy different criteria from ordinary, paradigm historical accounts. But then the difficulty is that philosophies of history resemble paradigm scientific theories hardly at all. If they resemble anything, they resemble paradigm historical accounts, except for their making claims on the future of a kind not commonly made by the latter.

This latter resemblance does not merely lie in the fact that, like historical accounts, philosophies of history often exhibit a narrative structure. It lies also in the fact that, typically, philosophies of history tend to give interpretations of sequences of happenings which is very like what one finds in history, and very unlike what one typically finds in science. Philosophies of history make use of a concept of interpretation which seems to me would be grossly inappropriate in science, namely a certain concept of 'meaning'. That is to say, they undertake to discover what, in a special and historically appropriate sense of the term, is the 'meaning' of this event or that. Professor Löwith offers the following as a general characterization of substantive philosophy of history. It is, he says, 'a systematic intepretation of universal history in accordance with a principle by which historical events and successions are unified and directed towards an ultimate meaning'.[1]

7

How are we to understand this special use of the word 'meaning', which is quite different from the way in which, for instance, we speak of the meaning of a term or of a sentence or an expression? Roughly, I think, as follows. We are to think of events as having a 'meaning' with reference to some larger temporal structure in which they are components. But this is not an altogether unfamiliar way of using this term. Think, for example, of the sort of critical point we score when we say, of a certain episode in a novel, or in a play, that it has no meaning, that it 'lacks significance'. We intend to say that it fails to further the action, that it is superfluous and hence aesthetically inappropriate. But this, of course, is a judgment we can make about a particular episode only once we have before us the entire novel, or only once the play is complete. Until then, we can say only that we do not know as yet what might be the meaning of the episode, though we assume it has some part to play in the progress of the plot. *Afterwards*, we might say that it had *this* meaning, or *that* (unless nothing, so to speak, turns with it,[1] that it was of no significance at all, a blemish in the well-made play). I emphasize that it is only retrospectively that we are entitled to say that an episode has a given specific meaning, and then only with respect to the total work. But information concerning the total work is just what we lack when we are going through it for the first time: then, if something strikes us as meaningless, we have to wait and see whether it is so; and if something seems to us to have a certain meaning, then again, we must wait and see if we are right. We are often obliged to revise our views concerning the meaning of an episode, in the light of what happens afterwards. This sense of meaning has application in history, too. Now that the French Revolution is over, we can say what was the significance of the Tennis Court Oath—something which even the participants in that event might have been wildly wrong about. So understood, we might think of philosophers of history as trying to see events as having meaning in the context of an historical whole which resembles an artistic whole, but, in this case, the whole in question is the whole of history, compassing past, present, and future. Unlike those of us who have the whole novel before us, and are able to say with some authority what is the significance of this event or that, the philosopher of history does not have before him the whole of history. He has at best a fragment—the whole past. But he thinks in terms of the whole of history, and seeks to discover what the structure of this whole must be like, solely on the basis of the fragment he already has, and at the same time seeks to say what is the meaning of parts of this fragment in the light of the whole structure which he has projected.

I quite agree with Professor Löwith's claim that this way of viewing the whole of history is essentially theological,[1] or that it has, at all events, structural features in common with theological readings of history, which is seen *in toto*, as bearing out some divine plan. It is, I think, instructive to recognize that Marx and Engels, although they were materialists and explicit atheists, were nevertheless inclined to regard history through essentially theological spectacles, as though they could perceive a divine plan, but not a divine being whose plan it was. Whatever the case, the substantive philosophies of history, insofar as I have correctly characterized them, are clearly concerned with what I shall term *prophecy*.[2] A prophecy is not merely a statement about the future, for a prediction is a statement about the future. It is a certain *kind* of statement about the future, and I shall say, pending a further analysis, that it is an *historical* statement about the future. The prophet is one who speaks about the future in a manner which is appropriate only to the past, or who speaks of the present in the light of a future treated as a *fait accompli*. A prophet treats the present in a perspective ordinarily available only to future historians,[3] to whom present events are past, and for whom the meaning of present events is discernible.

It is just here that I want to take up again my earlier claim that substantive philosophy of history is connected with history. We can now see how a philosophy of history resembles an ordinary historical account, for one thing. And we can understand how it sometimes happens that philosophies of history are even ascribed to the wrong genre, and taken merely as *very* ambitious instances of ordinary historical writing, on a specially grand scale: 'The difficulty with the grandiose proposals of the Marxes, the Spenglers, the Toynbees ... can hardly be that they are history, but that they are grandiose.'[4] The resemblance is due to the fact that philosophies of history make an unjustified use of the same concept of 'meaning' which has a *justified* application in ordinary historical work. I shall discuss later some of the problems which arise in connection with this notion of meaning, but for the moment it is enough to indicate how ordinary ascription of meaning to events is used in historical discussion. We might, for instance, know that what an individual *B* accomplished was due, in some considerable measure, to the influence upon him of *A*'s work. To demand, in an historical way, to know the significance of *A*'s work, is to expect some such answer as this: its significance is that it influenced the work of *B*. Obviously, this sense of 'significance' does not exhaust the whole meaning of the concept of significance: a corpus of poetry may be significant only because it is intrinsically great poetry. And it is perhaps arguable that unless we used the term 'significant' in some other, and non-historical sense, we would have no use at all for the historical sense. It may, that is, be true that we find *B*'s work intrinsically significant, a great achievement; and because of this, we are likely to regard the episode in *B*'s biography in which he first encountered *A*'s work as *fraught* with significance, indeed as laden with destiny. A contemporary would, of course, have been blind to this significance, for *B*'s great work had not been done. He would lack what we possess, namely the sort of information available only after this encounter. *Afterwards* a biographer might single this episode out as the most significant event in *B*'s life. A contemporary could not have seen it in this way: he might, indeed, regard it as beneath mention. *A*'s work, meanwhile, might have as its *sole* significance that it influenced the work of *B*.

Think, in this connection, of certain very common sorts of emotions which are connected with both memory and appraisal of our own actions and negligences, for instance, regret or remorse. Typically, conventionally, we express regret by saying 'If only I had known ...'. Now the ignorance we complain of here is often an ignorance of the future, an ignorance which has been removed by time, so that we now know, as we did not and perhaps could not then have known, what were the consequences of our actions or failures to act. Generally what we mean is that if we had known then what we know now, we would not have acted as we did. Such statements, of course, are puzzling. If, for instance, I know that *E* will happen, it follows that '*E* will happen' is true, so that *E* must happen. If *E* *must* happen, then nothing can be done to prevent its happening, or to make '*E* will happen' false. And so regret is gratuitous.

If, on the other hand, I can do something to forestall E, then it is not the case that E *must* happen. And if I do forestall E, 'E will happen' is false, and so I cannot be said to know that E will happen. If I can do something about the future, the future cannot be known; and if it can be known, we can do nothing about it. This is an old puzzle, of Aristotle's, and one we shall be obliged to reckon with later. But I am suggesting that 'If only I had known...' cannot be *strictly* taken: if I *had* known, I could have done nothing. To regret, however, presupposes that we do not see our own actions, at the time we perform them, as having the significance we will later attach to them, in the light of further events to which they are to be related. But this is a general insight into the historical organization of events: events are continually being re-described, and their significance re-evaluated in the light of later information. And because they have this information, historians can say things that witnesses and contemporaries could not justifiably have said.

To ask for the significance of an event, in the *historical* sense of the term, is to ask a question which can be answered only in the context of a *story*. The identical event will have a different significance in accordance with the story in which it is located or, in other words, in accordance with what different sets of *later* events it may be connected. Stories constitute the natural context in which events acquire historical significance, and there are a number of questions I cannot even touch upon at this point concerning the criteria belonging to a story, the criteria, that is, by appeal to which we say, with respect to a story S, that an event E is part of S and an event E' is not. But obviously, to tell a story is to exclude *some* happenings; is to appeal tacitly to some such criteria. Equally obviously, we can only tell the story in which E figures relevantly if we are aware of what later events E is related to, so there is a certain sense in which we can tell only *true* stories about the *past*. It is this sense which is somehow violated by substantive philosophies of history. Using just the same sense of significance as historians do, which presupposes that the events are set in a story, philosophers of history seek for the significance of events before the later events, in connection with which the former *acquire* significance, have happened. The pattern they project into the future is a narrative structure. They seek, in short, to tell the story before the story can properly be told. And the story they are interested in is, of course, the whole story, the story of history as a whole. This does not mean, to be sure, that every event is going to be part of the story (stories, to be stories, must leave things out), and this means, amongst other things, that the philosopher of history will be seeking for the significant events, the events that belong to the whole story. His mode of organization, then, is indeed the historical mode of organization. But the difference is not merely a certain grandioseness, as we shall see. It also has to do in an important way with a certain sort of claim on the future.

There are ways of finding out what will happen, and even ways of giving historical descriptions of things that will happen. One sure way of doing this is to wait and see what happens, and then write the history of it. But the philosopher of history is impatient. He wants to do now what ordinary historians, as a matter of course, will be able to do later. He wants to view the present and the past in the perspective of the future (indeed of the ultimate future, for there must be an end to every story). And he wishes to be able to describe events in a manner not ordinarily accessible at the time when the events themselves take place. There are descriptions, and I shall be much occupied with them in this work, which we encounter in history books, and which are cast in a mode quite characteristic of historical utterance—descriptions we find intelligible enough, and consider true, but which, with a suitable shift in tense, we would find very nearly unintelligible and hardly credible if they had been uttered at the time when the event they describe took place. An historian might write: 'The author of *Rameau's Nephew* was born in 1715.' But think how odd it would be were someone to have said, at the right moment in 1715, 'The author of *Rameau's Nephew* is just born'. Or even more puzzlingly, if someone were, in the future tense, to say the same thing in, say, 1700. What could such a statement mean to anyone in 1715, much less in 1700? One might, of course, predict that Mme Diderot would give birth to an author, even an encyclopaedist ('Unto you an encyclopaedist is born'), for instance, on the grounds that males in the Diderot family had been literary men for generations. But to refer, by title, to the potential author's unwritten works goes beyond prediction: it involves speaking in a prophetic vein, i.e. describing the present in the light of things which have not as yet happened ('Unto you a Saviour is born'). Yet it is just such descriptions of events, descriptions which make an essential reference to later events—events future to the time at which the description is given—that substantive philosophers of history undertake to give. In effect they are trying to write the history of what happens *before* it has happened, and to give accounts of the past based upon accounts of the future.

It is this about substantive philosophy of history that I find both philosophically interesting and philosophically odd. Critics sometimes draw upon an important distinction between the meaning *of* history, and meaning *in* history,[1] in order to cast doubt upon the legitimacy of the entire enterprise of philosophical history. To demand the meaning of an event is to be prepared to accept some context within which the event is considered significant. This is 'meaning in history', and it is legitimate to ask for such meanings. Usually, the context within which an event is significant is some limited set of events which may together constitute a whole in which the event in question is a part. Thus Petrarch's ascent of Mt Ventoux is significant within the set of events collectively making up the Renaissance (and perhaps not uniquely significant in that context). But we may also ask for the significance of the Renaissance itself. And this in turn requires specification of a larger context, etc. There are wider and narrower contexts, but history as a whole is plainly the widest possible such context, and to ask the meaning of the *whole* of history is to deprive oneself of the contextual frame within which such requests are intelligible. For there is no context wider than the whole of history in which the whole of history can be located. This is an important critical point, but not, I think, essentially damaging to the substantive philosopher of history. The philosopher might say that the whole of history gets its meaning from some quite non-historical context, for example, from some divine intention, and then go on to say that God at any rate is outside of history and indeed outside of time. Secondly, *he* might point out, as I already have done, that the ascription of historical significance is dependent upon the ascription of some other and non-historical kind of significance. For instance, *A* is significant historically for having in-fluenced *B*, because we regard *B*'s work as (perhaps) significant in some quite different sense. The philosopher might continue, then, by suggesting that we cannot speak of the historical significance of history as a whole,

13

but that historical significance is by no means the only kind of significance. Finally, he might emphasize that by 'history as a whole' he does not necessarily mean every event that *has* happened and every event that *will* happen. Perhaps not everything is part of history as a whole, nor is history as a whole *the* widest possible context. A story, we have said, must leave things out. Nothing that happened in Siberia, for instance, was considered by Hegel to be part of history.[1] He was not denying that things happened in Siberia, but only that these happenings had any significance in the grand march of events, the story of which it was his aim to tell. Discussing the meaning of history as a whole, he supposed it to be this: the progressive coming to self-consciousness of the Absolute. Each thing that happened in history was significant with respect to this story, or insignificant, but Hegel never asked what was the significance of the Absolute's final self-awareness. Or, if he had, he would doubtless have moved to a quite different sense of 'significant' than that applied to the ordinary events of history. Whatever mistake it is the philosopher of history is making, it is not, I think, the mere confusion of two senses of meaning. And even ordinary historians, as I have contended, could not always use 'significant' in just one way. If nothing were of non-historical interest, it would be meaningless to say, of something (like the eighteenth-century Neapolitan paintings) that they were of *merely* historical interest.

I feel nevertheless that substantive philosophy of history is a mis-conceived activity, and rests upon a basic mistake. It is a mistake, I shall argue, to suppose that we can write the history of events before the events themselves have happened. The error might be represented like this: it is an attempt such philosophers are making to give temporally inappropriate descriptions to events, to describe events in a manner in which they cannot be described at the time the attempt is made. I am appealing here to the familiar fact that we write the history of events after those events have happened. But of course, no such appeal constitutes an argument, and the proper philosophical question is why this fact holds, if indeed it holds at all. Scientists make unexceptionable claims on the future, as do all of us in practical life. But it is the *kind* of claim on the future which philosophers of history make, or which their enterprise requires them to make, which I find suspect. Their claims concerning the past and the present are, I maintain, logically connected with their claims on the future, so that if the latter are illegitimate, the former are not compelling. Historians describe some past events with reference to other events which are future to them, but past to the historian, while philosophers of history describe certain past events with reference to other events which are future both to these events and to the historian himself. And I wish to maintain that we cannot enjoy a cognitive standpoint which makes such an activity feasible. The mode of organizing events which is essential to history does not, I shall argue, admit of projection into the future, and in this sense the structures in accordance with which these organizations are effected are not like scientific theories. And this is, in part, due to the fact that historical significance is connected with non-historical significance, and this latter is something which varies with variations in the interests of human beings. The stories historians tell must not be relative merely to their temporal location, but also to the non-historical interests they have as human beings. There is, then, if I am right, an unexpungeable factor of convention and of arbitrariness in historical description, and this makes it exceedingly difficult, if not impossible, to speak, as the substantive philosopher of history wishes to, of *the* story of the whole of history, or, for that matter, *the* story of any set of events. Philosophy of history is an intellectual monster, a 'centaur', as Jabob Burkhardt once called it,[1] which is neither history nor science, though it resembles the one and makes claims for itself which only the other can make.

History co-ordinates, Burkhardt writes, and philosophy subordinates, and the expression 'philosophy of history' is a contradiction in terms.[2] This is true in a general kind of way, but it tells us very little about the way in which history co-ordinates which makes it so different, as we feel intuitively that it is different, from science. And this brings us to analytical philosophy of history, one main purpose of which is to clarify this mode of co-ordination. The main fact to keep in mind, for this purpose, is that the events co-ordinated are temporally distant from one another, that they are respectively past and future to each other, though both past to the historian. Why and whether they *must* both be past to the historian is the main question this book will deal with. So in discussing our knowledge of the past, I cannot but be interested in discussing our knowledge of the future, if we may speak at all of knowledge here. So in a way I shall be at least as much interested in substantive philosophy of history as I shall be in history itself. I shall maintain that our knowledge of the past is significantly limited by our ignorance of the future. The identification of limits is the general business of philosophy, and the identification of *this* limit is the special business of analytical philosophy of history as I understand it.

II

A MINIMAL CHARACTERIZATION OF
HISTORY

My thesis, in the last chapter, was that substantive philosophy of history belongs to a different genre from history itself, and that it consists in making projections, which I regard as illegitimate, into the future, of the same sorts of structures which historians employ in organizing the events of the past. It is because of structural similarities between ordinary historical accounts and philosophies of history, and again, because the same concept of historical *meaning* determines the kind of account appropriate here, namely, a narrative account, that we might be inclined to suppose that these two kinds of activities are generically of a piece, differing only in scope. An ordinary historical account covers only a piece of what a philosophy of history tries to cover, namely, the whole of history. Now of course there are differences of scope within history itself. A history of the Terror of 1793 has a narrower scope than the history of the French Revolution, and a history of the latter a narrower scope than a history of France, and this in turn a narrower scope than a history of Europe, and so on. The widest possible historical account, I have suggested, is an account of the whole past, an account which is to be distinguished from an account of the whole of history, the latter being an instance of a *philosophy* of history. It is tempting to suppose that there is only a practical impossibility in giving an account of the whole past, even if there is, perhaps, a logical impossibility in giving an account of the whole of history. But in fact this is not quite the case, and to see the reason for this is to see, in what I regard as the deepest sense, the manner in which substantive philosophy of history is 'connected' with history. I shall seek to argue, in a later chapter, that any account of the past is *essentially* incomplete. It is essentially incomplete, that is, if its completion would require the fulfilment of a condition which simply cannot be fulfilled. And my thesis will be that a complete account of the past would presuppose a complete account of the future.

so that one could not achieve a complete historical account without also achieving a philosophy of history. So that if there cannot be a legitimate philosophy of history, there cannot be a legitimate and complete historical account. Paraphrasing a famous result in logic, we cannot, in brief, *consistently* have a *complete* historical account. Our knowledge of the past, in other words, is limited by our knowledge (or ignorance) of the future. And this is the deeper connection between substantive philosophy of history and ordinary history. It is the reason why one cannot bypass substantive philosophy of history if one is analytically interested in the concept of history, even history as practised by ordinary historians.

A brief illustration will perhaps show what I mean. A complete account of an event would have to include every true historical description of that event. Consider the birth of Diderot in 1715. One true historical description of what took place is that, on that date in 1715, the author of *Rameau's Nephew* was born. Prior to the writing of *Rameau's Nephew* one could not so describe the event unless one made a certain kind of claim on the future, that is, without speaking in the prophetic mode. Such an historical description during the required interval would logically presuppose a sentence which belongs in the *philosophy* of history. But without this description, we do not have a *complete* description of the event in 1715. Hence we could not have a complete historical description without presupposing the achievement of a philosophy of history. And this is perfectly general. There will always be descriptions of events in 1715 which will depend upon descriptions of events which have not as yet happened. Only when they have happened can we give these descriptions, and without these descriptions, we have not a complete account. But to give these descriptions before the required events have happened is to do philosophy of history. So if philosophy of history is impossible, complete historical accounts are impossible as well, and historical accounts are thus *essentially* incomplete.

Notice that if philosophy of history were legitimate, a philosophy of history would licence, would indeed entail, certain statements about the past which historians would not otherwise be able to give. Philosophers of history do not *only* make statements about the future, they make statements also about the past. So 'making statements about the past' does not serve to distinguish historians from philosophers of history. This is the final connection I shall remark on between history and philosophy of history. The difference lies in what their respective statements about the past *presuppose*, and what marks the philosopher of history off is that his statements about the past presuppose certain statements about the future. By 'the future', of course, I mean 'his future'. The characterization of both history and philosophy of history essentially involves reference to the temporal location of the historian and the philosopher of history alike.

For the present, I shall be concerned only with ordinary history. Historians, *as* historians, are not concerned with events in *their* future, or at least not concerned with them in the way in which they are concerned with events in their past, or with, in certain cases, events in their present, events they are living through. They may be concerned with events they are living through in *this* sense, for example: they observe these events in the expectation that someday, when these events are past, they will write the history of them. This was the case with Thucydides, whose work is particularly instructive in our present context. He begins his celebrated book with the following sentence: 'Thucydides, an Athenian, wrote the history of the war between the Peloponnesians and the Athenians, beginning at the moment it broke out, and believing that it would be a great war, and more worthy of relation than any that had preceded it.'[1] It is plain that he felt that the set of events he was living through was 'significant', that there was to be an important story to tell, and he observed things as they happened, in order to be able afterwards to tell the story of them. Thucydides tries to be as accurate as it was possible for him to be in finding out what really had happened, an accuracy, he tells us, that cost him considerable labour.[2] For he was able to witness personally but a fraction of the events which belonged to his story, and was forced to depend, in the other cases, upon the reports of others. But these did not always agree, and in order to determine which (if either) of a pair of contrary accounts was correct, he was obliged to apply 'the most severe and detailed tests possible'. It is this care which earned for him the honour of being the father of scientific history. But the reason for his taking such pains was not *simply* to write a correct account, though writing a correct account was a necessary condition for what he further wished to do. He wished his work to be a *useful*[1] work, and he was persuaded that it could not be useful unless it was true. Hence his pains to be correct. Now perhaps all historians wish their works to be useful. But often the utility of their work has to do with nothing but the writing of more history. Their works, that is, are useful to other historians, interested in the periods or events covered, or to non-historians interested in finding out what happened. The criteria of utility remain within the area of history itself; but Thucydides was concerned with usefulness in a non-historical sense. He had, if you wish, a non-historical purpose in writing a work of history, as well as the obvious historical purpose. It is this further kind of utility that I want briefly to discuss.

His book was aimed at an audience which, in his words, 'desires an exact knowledge of the past as an aid to the interpretation of the future.' For 'the future must resemble the past if it does not reflect it'.[2] His work, then, is to be 'a possession for all time'. From these few statements, it would plainly appear that Thucydides is writing for what, according to my conception of them, must be philosophers of history: he is writing about the past, but only, or mainly, to provide a guide to the events of the future which 'must resemble if not reflect' it. But his reference to the future is *not*, I think, the kind of reference to the future which characterizes philosophies of history, and the claim that the future must 'resemble if not reflect' the past is, really, only a crude formulation of what we all recognize as the Principle of Induction. I offer the following, then, as a reconstruction of what Thucydides had in mind. Here, he is saying, is a war, taking place under these and those conditions. Conditions similar to these will hold in the future as they have held in the past, and so wars similar to this one will take place in the future as they have taken place in the past. Hence, if one can but determine what these conditions were in the present instance, then, if, in any future instance, we can identify similar conditions we can expect similar events to happen. And so we will have provided a guide to future events which resemble the Peloponnesian War, or resemble it sufficiently.[3]

Now this statement admits of a thoroughly trivializing interpretation. Of course we would be able to predict the sequence of happenings in every war which *sufficiently* resembles the Peloponnesian War, providing we had an accurate description of the Peloponnesian War itself: whoever understands *x* understands every copy of *x*, providing it is a good copy. The real question is whether in fact there *are* wars which sufficiently resemble the Peloponnesian War for us to do this, whether, indeed, the events of the future will really resemble 'if not reflect' the events of the past. And in a way we can say that they have not done this. We, for instance, have knowledge of a good many wars which are in our past but were in Thucydides' future, between which and the Peloponnesian War the dissimilarities are at least as striking as the similarities. To be sure, Thucydides could not have known about the wars in his future. But he must have had comparable information if we are to take him literally. For after all, the war he so brilliantly described was future to many wars in Thucydides' *past*, and presumably he knew about some of these. If his war *exactly* resembled those others, then we should have to reject many parts of his account. For instance, he makes a claim that his war was on a 'grand scale'. But he also tells us that 'the evidence which an inquiry carried back as far as was practicable leads me to trust, all point to the conclusion that there was nothing on a grand scale, either in war or other matters'.[1] Actually, we must suppose there was something altogether singular about his war if his statement that *his* war was 'more worthy of relation than any that had preceded it' is to be justified. If the future is to resemble the past, the past must resemble the future ('resembles' being a symmetrical relation), so it is plain that Thucydides cannot insist, as he does, on the unprecedentedness of his war, and also say that all future wars must resemble it. Obviously all wars in some sense resemble one another if they are to be so called. But then we need only consult diction- aries, or grammatical usage, to determine what are these common features in virtue of which an event is correctly to be called a war. We need not study history. In the cognitively important sense, however, what right had Thucydides to suppose that future wars would resemble his war, if past wars did not? Surely he could not have evidence for this, for the only evidence he had would have told strongly *against* the claim that future wars would sufficiently resemble this one. So how could he suppose other than that future wars would be as dissimilar to this war, as this war was dissimilar to all past wars? But if this was the only conclusion he was entitled to, given the evidence he possessed, his purpose in writing a 'useful' history was defeated from the start. Or, its utility would be just the reverse of what he thought it would be, namely, that the past is a wholly useless guide to the future, drawing the moral that if one really wants to cope with the future, one had better not waste one's time with history. Yet if the past is no guide to the future, what is? And surely in some sense, the sense enshrined in our inductive procedures, it is a guide. The question is whether Thucydides was entitled to an induction here.

There is something highly artificial in applying this sort of logical pressure to Thucydides' remarks.[1] Yet he is explicit about wanting to write a useful book, and explicit, as well, about what he thinks its use is to consist in. A more congenial reconstruction of his methodology might be this. As with Plato in a famous part of the *Republic*, Thucydides might have adopted the stratagem of taking something as another thing 'writ large'. The idea was that if A is a magnified projection of B, then the structural features common to A and B might be more readily studied in A than in B. It is, of course, presupposed that A and B *are* structurally alike, but some such presupposition lies behind our unquestioning acceptance of microscopy: even if we are unable with the 'naked eye' to compare an enlarged image of x with x itself. So Thucydides may have felt that the war which was taking place was on so grand a scale that one might regard it as a magnified instance of the whole class of wars, and so, by studying it, one could discern structural features not so readily made out in smaller instances. He says after all, or strongly suggests, that it was its *size* that made this war 'more worthy of relation' than any other. His selection of it then might be justified in the way in which selection of a particularly clear instance of anything is justified if other members of the population from which it is drawn exhibit the characteristic features of the class less lucidly than it does. His narration is meant to bring out the features of typical human responses to typical situations which he believed recur again and again. And it is in just such terms that his work has been appreciated since, and has been considered, as he wished it to be, a 'possession for all time'. and as something more than *merely* an account of what happened between Athens and Sparta in the long dead past.

We might then say that his claim that the future must resemble the past, and that by describing clearly a splendid instance of the class of wars, he was furnishing a guide to future wars, was not an *essential* temporal claim. He is making no explicit temporal reference in speaking of the future. He could, with equal justification have said that he was providing, in the appropriate sense, a guide to all *past* wars. But the latter he would not perhaps have regarded as a worthwhile contribution. A practical man, Thucydides must doubtless have felt that nothing could usefully be done about the past (as Richard Taylor has suggested, we are all fatalistic about the past).[1] It is only the future we can do anything about, and it is only with respect to the future that his work would be likely to have its intended use.[2] But this is in fact immaterial so far as the *logic* of his argument is concerned in our present interpretation. For what in effect he was doing was to argue from a sample (albeit a 'good' sample) to a population—from the current war, to *all* wars, past *and* future. But 'past *and* future' adds nothing to the expression 'all wars'. Thucydides was then making a claim on the future which is like the claim that any of us make in performing any standard induction. It is a claim, however, which it is misleading to think of as any more a claim on the future than it is a claim on the past. For it is, rather, a claim concerning a population, and is independant of any information concerning the temporal location of its members, either in relation to one another, or in relation to the person who makes the induction. It is true that we sometimes phrase our philosophical doubts about induction by asking 'Will the future be like the past?' But there is in fact no temporal direction in induction, which is symmetrical with regard to time. And we could as easily phrase the question thus: 'Will the *past* have been like the past?' That is, will the parts of the past that preceded the period of the past from which our samples have been drawn 'be like' the latter period? For all the same problems regarding future examples of the population from which my sample has been drawn are raised by examples of the same population which were temporally earlier than the drawn ones. For instance, we have no better grounds for supposing that there will even *be* later instances than we have for supposing that there ever *have been* earlier instances. And just as there is a possibility that the samples we have are the *last* ones, there is a matching possibility that those we have are the *first* ones. Hume once entertained the logical possibility that the whole complexion of the world might someday change, after which none of our general laws would hold.[3] But we can entertain the matching possibility that this has already happened, that such a change may have taken place, and that the world was once as dissimilar to what it is as, on Hume's supposition, it will be. Clearly, I have no inductive grounds for ruling out either possibility. For it is precisely the limitations of such grounds that Hume's possibility was meant to illuminate. Or rather, I have only inductive grounds for ruling out either possibility, since each is logically coherent: and inductive grounds are inadequate here. To suppose them adequate is to beg the question.

Without protracting this schedule of symmetries, we may conclude that inductive processes are invariant as to the direction, pastwards or futurewards, of inferences to unexamined instances. The implicit conclusions, then, of Thucydides' work, that human motivations are everywhere and always the same, that humans respond in predictably standard ways to predictably standard situations, are time-independent. Hence he makes no different sort of claim about the future than he makes about the present or the past. I am not concerned, of course, to question whether his specific conclusions are right or wrong, but only to insist upon their time-independence. So Thucydides is not engaged in philosophy of history, as I have characterized it. Instead, he is writing social science, at least implicitly, for he intends that we should gather from his work some very general facts concerning the behaviour of individuals and groups in political contexts, facts which are particularly well illustrated in the events he is narrating. But of course his work would remain valuable even if these were not general facts, even if, indeed, the Greeks and Spartans were significantly different from those who came later or earlier. We in fact see ourselves mirrored in his book.

The success of Thucydides' illustration depended, apparently (at least from his own point of view), upon his giving as accurate an account as he could of what actually happened. So the very least we can say about Thucydides is that he endeavoured to give a true description of events in *his* past, some of which he had witnessed and some of which he had not, but which had been witnessed by contemporaries of his whose testimony he submitted to the most stringent probation. And this could be said of him even if he had not had the ulterior purpose I have been discussing. Allowing that this ulterior purpose may have dominated his selection of what features of the war to remark upon, we may nevertheless consider the two activities as in some sense independent, and distinguish those features of the same work which satisfy the complementary descriptions 'is a work of history' and 'is a piece of social science'.

I shall employ Thucydides' stratagem of making a general point concerning a whole class through exhibiting a good instance of that class, by taking *him* as an especially good instance of the class of historians. I shall say that the very least that historians do is to try to make true statements, or to give true descriptions, of events in *their* past. I offer this as a minimal characterization of historical activity, as a necessary condition for applying the predicate 'is an historian' to an individual. I do not say that it is a sufficient condition, for, as we have seen, it is also part of our criterion for applying the predicate 'is a philosopher of history' to an individual. Perhaps we can amplify this criterion to make at least the distinction between historians and philosophers of history by saying that historians, in contrast with philosophers of history, try to make true statements, or to give true descriptions, of events in their past which do *not* logically presuppose true, and time-dependent, statements about, or descriptions of, events in *their* future.

I am not saying that this is all that historians do. But I will insist that *whatever else* it might be that historians are thought of as trying to do, their success in making such statements is a necessary pre-requisite for that other activity, whatever it might be. Thus it might be said that historians seek to *explain* events in their past. I cannot quarrel with this. I only say that it is first necessary to give a true description of the event to be explained. But what if an historian *A* has already given such a description of an event which an individual *B* wishes to explain? Would we then call *B* an historian? The answer is that any explanation of an event will require reference to another event, and unless we have a true description of it, we shall not have succeeded in explaining the given event. Nor ought we to overlook that '*E-2* happened because of *E-1*'—supposing this to be offered as an explanation of *E-2*—is, at least, a true statement about some event in some historian's past. Similarly, if it is said that in order to explain some past event, an historian must perform a special act of empathic identification with persons involved in that event. I have no doubt that historians can and do perform such acts. But surely their ability to do this non-vacuously depends upon their first having

25

established that there was such an event, and that there was such a person with whom empathic identification can be attempted. And *this* cannot be achieved by empathic identification. This, however, suggests a gap in my characterization. For it might be argued that a person is not an event, and that my characterization only has to do with events in the historian's past. So I shall amend my characterization to compass making true statements about the past, whether these be about events, or persons, or things of any kind.

This is all, for the moment, that I wish to say about historians. I do not even want to insist that they ever succeed in what they are said to be trying to do, but *only* that they try to do this. And surely this is an innocuous enough thing to say about historians. It is perhaps not even very enlightening to say this, not, at least, until we give some further particulars of the kind of statement they try to make (a true statement is not, in this sense, a *kind* of statement). We might similarly say that a philosopher of history tries to make a statement of a certain kind about the future. But what I wish to say is that what the substantive philosopher of history attempts is to make the same kind of statement about the future that historians try to make about the past. So our picture of the substantive philosopher of history will acquire shape as we get a better picture of the historian himself. And we will, I hope, eventually be able to see why it is illegitimate to make, about the future, the kind of statements it is legitimate to make about the past.

Meanwhile, there is some reason for keeping our characterization of the historian's intention as general and unspecific as I have. The reason is this. It is sometimes argued that we cannot *in general* succeed in making true statements about the past. But if, in general, the historian's intentions are unfulfillable, there is little to be gained by any further descriptions of the intention. If there are no unicorns, it is idle to ask for details about unicorns, for example, whether they are savage or mild. I turn therefore to the objections there might be against our making true statements about our past.

III

THREE OBJECTIONS AGAINST THE POSSIBILITY OF HISTORICAL KNOWLEDGE

Few of us, I think, have any serious doubt that historians sometimes succeed in achieving that aim which I have minimally ascribed to them, that they sometimes, indeed frequently and typically, succeed in making true statements about things in their past. The question is whether we are justified in supposing this. To raise such a question, of course, is not to cast doubts on the competence or integrity of historians. We plainly have ways of discerning incompetence or mendacity, and are usually able enough to determine whether historiographical skills are being abused or misused. The question, rather, is whether these skills enable us to achieve the minimal purpose for the sake of which we take the trouble to master them, and allow us to make true statements about things in our past, or to decide whether any statement which purports to do this is true or false. The question is more general even than this. For suppose it could be shown that the skills, the mastery and honest employment of which qualify someone, by present criteria, as an historian, were somehow utterly insufficient for achieving our minimal purpose. It is hardly plausible to suppose this could be shown, but if it were, men might then undertake to find another set of skills better fitted for achieving this purpose than the present ones. Surely it has happened, in the history of thought, that a set of techniques, imagined to be sufficient for attaining a given end, for example, for solving a certain kind of problem, were shown not to be so, so that new and more powerful techniques had to be found. But I am not concerned here with objections against presently accepted historiographical skills. I am, rather, concerned with objections against our being able, with *any* set of techniques, to make true statements about the past, so that further improvements in existing techniques would be as idle, say, as further improvements upon existing compasses would be once it were demonstrated that one cannot, by means of ruler and compass alone, trisect an angle. To put the question in this *general* form is to mount an attack upon the foundations of historical knowledge. And it is this attack which is to concern me now.

It does not commonly occur to one to adopt a position of wholesale scepticism with regard to statements purported to be about the past. One may doubt this statement or that, but usually for some fair reason, for instance, that one distrusts the person who makes it, or finds the evidence offered in support of it in some way defective, or rejects it because it conflicts with some other statement we are prepared to trust. Often, in fact, that other statement will *itself* be a statement about the past. Thus we might reject the statement that Sir Walter Raleigh was an atheist because we accept, as true, certain other statements about Sir Walter's behaviour which are incompatible with his having been an atheist. And in such a case, we are always at least prepared to accept the natural contradictory of a rejected statement, that is, that Sir Walter was *not* an atheist—itself a statement about the past. We can accept a wholesale scepticism here only if the acceptance of *any* statement purportedly about the past conflicts with some other statement which we are prepared to accept as true, and which rules out *any* statement about the past; that rules out *both* 'Sir Walter Raleigh was an atheist' and its natural contradictory. But any such proposition must be wholly general if it is to justify *wholesale* scepticism, that is, if it is to entail the unacceptability of both *p* and not-*p*, if *p* is a statement purportedly about the past. By the *natural contradictory* of a statement, I shall mean a contradictory which preserves the same subject, predicate, and tense, of the rejected proposition. So that '*S* was not *P*' is the natural contradictory of '*S* was *P*'.

I shall now state briefly three distinct arguments which, if cogent, entail the impossibility of making any true statement about the past, and which justify a wholesale scepticism towards both *p* and not-*p* if these are in the past tense. These arguments attack statements purportedly about the past at three different points: their meaning, their reference, and their truth-values. I do not believe any of these arguments in fact to be coercive. It is, moreover, easy to see what, in general, is wrong with each. But to work through each of them in detail is not merely philosophically instructive, since the arguments themselves are philosophically interesting. It will, in addition, bring out different aspects of the concept of history, and it is this which will justify, I hope, the rather extended treatment I propose to give these arguments in later chapters. For the present I shall merely state, and briefly comment upon, each one.

(1) Every statement purportedly about the past is strictly speaking *meaningless*. But then, with meaningless statements, the question whether they are true or false cannot, in principle, arise. So, if we cannot make a meaningful statement about the past, we cannot make a true statement about the past.

Now this argument presupposes a certain theory of meaning. The sophisticated reader will recognize, in fact, that what is presupposed is the celebrated Verifiability Criterion of meaning, which, on one of its many formulations, holds that a non-analytical proposition is meaningful only where it is verifiable by experience. Sometimes this was taken to imply that we must be able to experience what such a proposition is about. But we cannot now experience what statements about the past are purportedly about, hence we cannot verify them, and hence, by application of the criterion, they are meaningless. Few are so puritanical or so heroic as to maintain this extreme view, least of all those framers of the Verifiability Criterion whose aim, after all, was not the extirpation, but the explication of empirical science. A moderated version, however, which holds that the meaning of an empirical sentence *is* just its mode of verification has consequences which are nearly as paradoxical. For amongst the modes of verifying historical statements we can hardly reckon experiencing what they are about. For we cannot now do that. What we do instead is to seek for evidence in support of them, and this then suggests that the meaning of an historical statement is the process of finding historical evidence, and that historical statements, accordingly, can be interpreted as predictions concerning the results of historiographical procedures. But all such procedures must take place after the pronouncement of the given historical statements whose meaning they are, that is, in the historian's *future*. And inasmuch as the meaning of a proposition is what a proposition is about, historical statements, when meaningful, are about the future. So we remain unable to make meaningful statements about the past. And so we have just the same heroic position as before. Notice that even from the more enlightened point of view of meaning, for instance, that the meaning of a sentence is its *use*, we would have roughly the same consequence. For it is the *use* of predictions to make statements about the *future*, and so, once more, we fail to be able to use historical statements to make statements about the past. The thesis that historical statements are (covert) predictions has been subscribed to, in various ways, by Pragmatists such as Peirce, Dewey, and Lewis; and by Positivists, in particular A. J. Ayer.[1]

(2) Perhaps argument (1) confuses meaning with reference, a not uncommon philosophical lapse. But here a different difficulty arises. For perhaps there is, or rather was, nothing for statements, purportedly about the past, to be *about*. At least it is logically possible that the world was created just five minutes ago, intact with us and all our memories, and containing all those bits and pieces of things we take as evidence for a much older world than we in fact inhabit. The whole present complexion of the world might be just as it is, independently of when the world was created, and the world, as we now know it, is compatible with an astoundingly brief history of itself. But then, if it were created five minutes ago, there would have been nothing for statements purportedly about the past to refer to. Hence depending upon which of the more favoured current analyses of what are called 'referring expressions' all such statements would be *false* (Russell) or else the question of their truth or falsity could not arise (Strawson).[2] But then with neither of these analyses could the minimal historical aim of making *true* statements about the past be achieved. Most historical disagreements would be spurious. For strictly speaking, each of a pair of disputing historians would be either asserting a *false* proposition, or else asserting a proposition about which the question of truth and falsity could not arise. But this is precisely the same as being sceptical about *p* and its natural contradictory, when *p* is a statement purportedly about the past.[3]

This argument, it must be noticed, is not strictly general, and hence involves a less wholesale objection against my characterization than does (1). For even if we allow that the world did come into being, intact, and so on, just five minutes ago, we could nevertheless succeed in making *some* true statements about the past, namely that the world came into being five minutes ago, as well as further statements about happenings within the past (indeed within the *only*) five minutes. The argument could not rule out every statement about the past because it of course presupposes at least one statement about the past in its own formulation.

Nevertheless, it permits so few genuine statements about the past that its failure to be perfectly general bears but cold comfort to the historical industry. For how many historians, after all, are concerned with what has happened only within the last five minutes?

The argument does not require, of course, that the world in fact began five minutes ago, but only that it *might* have 'for all we know'. It might or might not have. So perhaps we can succeed in making true statements about the past, and perhaps we cannot. If we succeed, we cannot *know* we have done so. For all the evidence is compatible with the world having come into being five minutes ago, and we have, then, no way of knowing, on the basis of *evidence*, whether we have succeeded or not. We are, then, never in a position to know whether our historical disagreements are genuine or not. But this then, is the same as being sceptical about p and not-p, when p is purportedly about the past. For when we are not in a position, and cannot in principle be in a position, to say whether or not a given proposition is true or false (or neither), what is this but scepticism with regard to that proposition?

In comparison with (1), few people have taken this argument seriously, except Bertrand Russell, who formulated it, and *he* said that no one could seriously maintain it. Nevertheless, it raises in a dramatic way a variety of questions about time, reference, and knowledge, and merits a careful examination.

(3) Historical statements are made by historians, and historians have motives for making historical statements about one past thing rather than another. Not merely that, but historians have certain feelings about the past things they are concerned to describe. Some of these feelings may be personal, some may be shared by members of various groups the historian belongs to. Such attitudes induce historians to make emphases, to overlook certain things, indeed to distort. Because of the baggage of attitudes they bring with them, they themselves are not always able to detect the distortions they make. But those who pretend to detect distortions have themselves a special set of attitudes, and hence their own manner of emphasizing, overlooking and distorting. Not to have attitudes is not to be a human being, but historians are human beings, and cannot, accordingly, make perfectly objective statements about the past. Every historical statement, as a consequence of unexpungeable personal factors,

is a distortion, and hence not quite true. So we cannot succeed in making statements about the past which *are* quite true.

This argument would seem, on the face of it, open to an easy charge of meaninglessness. What, for instance, would it mean to say that every object in the world were crooked? We can only determine crooked things in comparison and contrast with straight things, and if there are no straight things, we cannot significantly apply the expression 'crooked'. It is a term which logically requires its polar opposite. But so with distortions. If we have no idea of what an undistorted statement about the past is like, what sense can we give to the expression 'distorted statement'? And if we *do* have such an idea, then we can in principle produce instances of undistorted statements, and the argument is wrong. So, this objection concludes, either the argument is meaningless or wrong.

But in fact this objection is not especially compelling, and the proponents of (3) can, and commonly do get around it easily. For they are not saying, in effect, something like 'Everything is crooked' but only that a certain class of things are crooked. Then there might be a class of straight things which would make this statement intelligible. So again, they are not saying that every statement is a distortion, but only that *historical* statements are. The class of historical statements is then contrasted as a whole with another class of statements, presumably undistorted, namely, the class of scientific statements. What Margaret Macdonald says about criticism in the following quotation can be applied readily enough to history:

Critical talk about a work is a construction of it by someone at a particular time, in a certain social context. Thus criticism does not, and cannot, have the impersonal character of strict rules, applicable independently of time and place, appropriate to science and mathematics.[1]

So we apparently know what sorts of accounts are 'objective', namely those which are independent of the time, place, and personal attitudes of him who gives it. But the precise criteria which enable us to know when an account is objective, enable us, as well, to know when an account is not. We ourselves cannot give an account of the *same kind* as an account we are claiming to be unobjective, which is itself objective. For any such account would again be relative to our own time, place, and personal attitudes. For we know that *any* account of that kind fails to be objective. And historical accounts are all of that kind.

Argument (3), in one form or another, has been defended by a number of thinkers of otherwise different persuasion. Nietzsche, for instance, used it in a celebrated aphorism, which was later approvingly cited by Freud. It runs: 'My memory said I did this. My pride tells me I could not have done this. My memory succumbs, and my pride remains inexorable.'[1] Pride here has distorted memory, and what I want to believe about the past distorts the truth. But it is of course logically possible that each of my memories has been warped by pride, or at all events by my attitudes or desires or feelings. So each memory *may* be a distortion 'for all that I know'. I have no way of knowing, that is, whether my memory is correct or not. So even if it is correct, I have no way of telling that it is. It may be objected that surely I have ways. I can appeal to independent evidence. But if this independent evidence consists in appeal to the memories of others, what grounds have I for supposing their memories to be any less distorted than mine? True, there is evidence of other sorts, for example, diary inscriptions, newspaper clippings, and the like. But just at this point the *general* relativistic argument (3) supervenes, and my assessment of evidence will again be influenced by personal factors, and so on. Nietzsche's argument, after all, is not restricted to memory. It may say about me, in my diary, that I did this or that. I disapprove of myself doing that, and my faith in the diary collapses: I say that someone else must have written it, or that I only did so to be clever.

This argument seems to me the most impressive of the three, despite the fact that their statement of it, by its main supporters, Beard, Becker, Croce, has been distorted by the special attitudes, prejudices, and feelings they had. It needs a good deal of logical scrubbing and polishing, but in the end there is something correct and important in it, and I shall subsequently modify my minimal characterization of history in terms of it. Indeed, I have already committed myself to views it must find congenial. For I have said that historical significance is dependent upon non-historical significance, and that the latter is very much a matter of the local attitudes and interests of the historian. So it will follow that our entire mode of organizing the past is causally involved with our own local interests, whatever they may be.

But I now shall consider all these arguments in the order in which I have stated them, and devote a chapter to each.

IV

VERIFICATION, VERIFIABILITY
AND TENSED SENTENCES

I now wish to consider argument (1). I shall approach it from the point of view of two distinct theories, each of which either entails it or provides it with a certain measure of philosophical support. The first is a theory of knowledge, and the second a theory of meaning. The two theories are, of course, importantly interconnected, and whoever subscribes to the theory of knowledge is likely also to accept, in one form or another, the theory of meaning, and conversely. It is nevertheless worth our while to consider each theory separately, for each of them serves to illuminate a somewhat different aspect of the concept of history. And though the points I shall be interested in making are capable of a perfectly general formulation, I shall illustrate the theories with reference to the work of individual philosophers who have, at one time or another, seen fit to defend them. The theory of knowledge I shall discuss is due to C. I. Lewis, and the theory of meaning to A. J. Ayer. Indeed, I shall examine *several* theories of Ayer's, concerned with the same problem, but reflecting certain changes in his basic philosophical programme.

A great deal of careful and important philosophical work has been done on the analysis of empirical knowledge since Lewis wrote *Mind and the World Order* in 1929, some of it by Lewis himself in his later and major work, *The Analysis of Knowledge and Valuation*. No one today, I think, would subscribe to the form of empiricism elaborated in that earlier work without some very severe qualifications. I shall nevertheless restrict my attention to what Lewis says in *Mind and the World Order*, for subsequent refinements do not bear significantly upon the problem which concerns me, and because Lewis says a number of interesting things in it which have to do with our knowledge of the past.[1]

I begin with a summary statement of the general theory of knowledge Lewis proposes, and shall then proceed to its specific application to our knowledge of the past. Lewis is in general concerned with claims to know that something *x* has a certain property *F*, and he maintains that when we in fact claim that *x* is *F*, we are to be understood as meaning something about actions and experiences, and it is in terms of actions and experiences that he undertakes to analyse sentences of the form '*x* is *F*.' He writes:

To ascribe an objective quality to a thing means implicitly the prediction that if I act in certain ways, specifiable experiences will eventuate: if I should bite this, it would taste sweet; if I should pinch it, it would feel moderately soft; if I should eat it, it would digest and not poison me; if I should turn it over, I should perceive another rounded surface much like this. . . . These and a hundred other hypothetical propositions constitute my knowledge of the apple in my hand. . . .[1]

In general,

The whole content of our knowledge of reality is the truth of such 'If–then' propositions, in which the hypothesis is something we conceive could be made true by our mode of acting and the consequent presents a content of experience which, though not actual now and perhaps not to become actual, is a possible experience connected with the present.[2]

Very roughly, then, the tense and common use and grammatical form of it notwithstanding, a sentence of the form '*x* is *F*' is a prediction, or better, a set of predictions of the form 'If *A* then *E*' where *A* marks an action and *E* marks an experience. And the original sentence is to be analysed[3] into these conditional sentences, a conjunction of which expresses our knowledge of what the original sentence claimed. Each of these conditionals states a separate process of verification, and the original sentence is *exhaustively* verified when all the conditionals into which it is analysable have been made true through the performance of the specified action, and the having of the specified experience. This is not an unfamiliar sort of analysis of the concept of empirical knowledge, and there are a great many problems which arise in connection with it, but I shall ignore them, and restrict myself exclusively to that part of Lewis's analysis which holds that when I claim to know something, I am implicitly predicting what I will experience if I do something, and that predictions concerning actions and their experiential outcomes represent the 'whole content of our knowledge of reality'.

Suppose, now, that I say, of a particular object *a*, that *a* is *F*, and that the sentence '*a* is *F*' is uttered at time *t*-1. If *a* exists at *t*-1, I may proceed to act upon *a*, and, depending upon the experiences I have as a result of this action, verify, or partially verify, my original sentence. Or suppose that I put my sentence in the future tense, and *a* exists after *t*-1. Here again, I will be able to perform actions, and have experiences, and so again will be able to verify, or partially verify, or even falsify my original sentence. For in either case, my original statement was a prediction of what I would do and experience. Now suppose *a* existed before the utterance of my sentence, and no longer exists, and that my sentence is in the past tense. Well, I cannot at *t*-1 act upon *a*, nor can I at any future time act upon *a*—things do not exist, stop existing, and come into existence again in the way in which they go from red to green to red again. And I cannot hope, in the future, to occupy that part of time occupied by *a*; so I have no way of verifying my sentence. To be sure, I might already have performed that action on *a*, and have had that sort of experience, which, had *a* still existed, I could respectively perform and experience; so that I would already have verified the sentence I utter in the past tense at *t*-1, the verification having preceded the sentence it verifies. But in fact the claim that I did perform an action, that I did have an experience, is in the past tense, and just the same problems would arise. So with regard to statements purportedly about the past, these preliminary considerations suggest that they cannot be verified, and therefore are not part of our knowledge of reality. Anyone can see his way through this argument, and it doubtless has an artificial ring to it, but Lewis regarded this as an objection which had some paradoxical consequences, and had to be met:

Knowledge, it is said, is here identified with verification, and verification comes about by proceeding from the present to the future. Then the past, so far as it can be known, is transformed into something present and future, and we are presented with the alternatives, equally impossible, that the past cannot be known or that it really is not past.[1]

Let us now see how Lewis himself handles this objection. He first of all denies that it applies, and insists that sentences about the past are verifiable and that we can know the past after all. But he supports this claim by introducing a novel conception of an *object*, and by permitting himself a variety of metaphysical assumptions which it is exceedingly difficult to justify by his account of knowledge. He says, to begin with, that

Verification

The assumption that the past is verifiable means that at any date after the happening of an event, there is always something which at least is conceivably possible of experience, by means of which it can be known.

Surely this is innocuous enough. Lewis is saying that our knowledge of the past is based upon present evidence, upon things we can in fact experience. These things he terms the 'effects' of the event we may claim to know on the basis of them. Should there be an event with no effects at all, or with no present effects, then, of course, we have no way of knowing that it happened: there would be a permanent hole in our knowledge of the past. This is roughly the solution to the problem which Dewey offers:

The object [of historical knowledge] is some past event in its connection with present and future consequences and effects.[1]

And again:

If perchance the past event had no discoverable consequences or our thought of it can work out no assignable difference anywhere, then there is no possibility of genuine judgement.[2]

Who could quarrel with this? His statement can be reduced to the claim that what can be known only on the basis of evidence cannot be known if we do not have the required evidence. And it assumes, what is surely not controversial, that we can only know about past events on the basis of evidence. Yet these genial platitudes must not still the unrest Lewis's analysis has left us with. For, his claim was that in saying something about the past, I am only predicting what experiences I will have in performing certain actions, and that the whole of my knowledge consists of these conditional propositions. So unless we *mean* by 'The Battle of Hastings', for example, some set of actions and experiences which lies in our own future, and it is unreasonable that we should mean this, what sense can we give to the expression 'Knowing that the Battle of Hastings took place in 1066' if the *whole* of our knowledge consists in a set of conditional sentences which refer to future actions and experiences? How, on such analysis, *can* I know the past, or anything other than these conditional sentences? And moreover, if I have no way of referring to past events, if, each time I try to refer to a past event, I find myself instead making a prediction about my future experiences, how am I to describe these

37

experiences as standing in some evidential relationship to a past event? For, instead, the moment I try to refer to the past event, I must be making a prediction of future experiences. How can I say that these experiences are evidence for p, where p ostensibly refers to a past event, when p itself is just a prediction of future experiences?

Lewis must have been dimly troubled by these problems, for it is just here that he introduces the novel conception of an *object* to which I referred a moment ago. Let E be an event, and let $\{e\}$ be the set of its effects at a given time t. Then, Lewis suggests, we may regard E, together with $\{e\}$, as one single time-spread object, stretching from the date of E, say t-1, to t. Presumably this object, which I shall call O, will go on growing in the direction of time, and fatten as new effects become part of itself. Thus the Battle of Hastings, plus the Bayeux Tapestry, plus all the other effects of the Battle of Hastings, go to form a single time-spread object. Let this be O. And since, as he has said, 'at every date after the happening . of an event, there is always something ... possible of experience ... ', there is at this very moment something, which is an effect of the Battle of Hastings, which it is temporally possible for me to experience. Hence I can experience O. To be sure, I could just *call* O 'The Battle of Hastings' and so say that I can experience the Battle of Hastings. But it would surely startle students of English history to learn that the Battle of Hastings is still going on. Nor would they be especially comforted by my so changing the meaning (the reference) of 'the Battle of Hastings' as to support this claim. It would be silly to say I saw Abraham Lincoln this morning, if all I meant was that I saw a copy of the Gettysburg address. So, strictly speaking, the introduction of O does not help. The fact that I can experience O does not mean that I can experience the Battle of Hastings. It only means that I can experience parts of the same temporal object of which the Battle of Hastings is a temporally earlier part. And since I cannot now experience temporally earlier parts of time-spread objects, we are left just where we were. We have only redescribed the problem by shifting from the question of our knowledge of past events to the question of our knowledge of temporally earlier parts of temporally-extended objects, when only present and future parts of such objects are capable of experience. Lewis writes that 'the totality of such effects quite obviously constitutes all of the object which is knowable'.[1] But this is precisely to say that the Battle of Hastings, not being one of its own effects, is not knowable. Not merely this. All temporally earlier parts of O are unknowable if the originating event is unknowable. And if we have ways of knowing *them*, why not ways of knowing the originating event of which they are the effects? Only present and future effects are knowable, and we remain unenlightened with regard to the question of how we know the past. Or rather, we are enlightened. For the answer is that we cannot. And this is absurd.

One word more. Suppose we experience {*e*} and {*e*} is indeed part of O. With what justification can we claim to know that {*e*} is part of O if O contains temporally earlier parts which are themselves unknowable? And if each time we wish to speak of temporally earlier parts, our statement turns out to be a prediction about temporally later parts, Lewis has in effect left no room in his theory of knowledge for the sorts of knowledge which reference to time-extended objects requires. Such reference in fact is impossible in this account of knowledge. But it is instructive to note the manner in which Lewis endeavours to circumvent these difficulties. He speaks, for instance, of 'marks of pastness' which presumably present objects bear, and on the basis of which we can find our way back to the temporally earlier parts of temporally extended objects they themselves are parts of. Thus:

The past is known through a correct interpretation of something given, including certain given characters which are marks of pastness.[1]

How are we to understand the expression 'marks of pastness'? As having reference to nicks, scratches, scuffs, and general signs of wear and tear? Or to date-inscriptions? Or simply differences from objects which bear marks of presentness? And what would *they* be? Lewis is remarkably evasive here:

For the present purposes it will be sufficient to remark that obviously *some* kind of identifiable marks must mean the pastness of the thing presented, otherwise the past could not be distinguished from the present.[2]

The question, however, is whether on Lewis's theory, we are able to do this. One is reminded here of a comparable puzzle in the empiricist theory of memory. By what present and identifiable criteria are we to distinguish memories from images, if we suppose that having a memory is to entertain an image? Hume suggested that we do so on the basis of some differential of vivacity,[1] but it has been pointed out that we can distinguish memories themselves by varying degrees of vivacity,[2] and the problem remains of distinguishing dull images from bright memories. Russell suggested that there is a certain 'feeling of pastness' which marks the difference,[3] and this is remarkably like Lewis's 'marks of pastness'. I do not intend going into the empiricist theory of memory here, but part at least of the difficulty surely lies in the assumption that memory consists in the beholding of an image, just as, in Lewis's case, knowledge is identified with a present experience, or an experience which will sometime be present. Hence the only way to account for our knowledge of the past is to look for some present mark of pastness.

I have no idea what a mark of pastness is, but if I were a forger of, say, Etruscan artifacts, I should want to make sure my nefarious confections exhibited sufficient of them so that gullible curators could not tell them from the real thing by 'marks of pastness'. And in plain fact we don't tell forgeries from real things by noticing that the latter are, in the words of one writer, 'stippled o'er with wasness'.[4] Instead, we have recourse to differential amounts of manganese and bitumin, to the presence of air-vents, and to the knowledge of the behaviour of terra-cotta under heat.[5] But at all events the issue is irrelevant to the present discussion. The relevant question has to do with taking something presently experienced as evidence for something which is past. 'Is evidence for' is a two-term predicate, and the question here has to do with the other term of the relation. For if we cannot refer to what it is that something is taken as evidence for, it is difficult to see how we can speak of evidence at all. And Lewis's inability to allow reference to the past which does not collapse into reference to present and future experience leaves us without a means for describing something even as evidence. Evidence for *what*? This we cannot say.

The criticism that the past is unknowable, which is entailed by his theory of knowledge, would have, Lewis tells us, 'greater weight if in general those who urge it were prepared to tell us how the past, which is really dead and gone, can be known'.[6] It is premature, perhaps, to rise to this challenge, but a few preliminary remarks can hardly be avoided.

Verification

Let us suppose that if E happens at t-1, then no one after that time can experience E. This is clearly presupposed in Lewis's discussion, and raises all those difficulties that such *ad hoc* remedies as time-extended objects and 'marks of pastness' are introduced, unavailingly, as we saw, to mitigate. Now someone is sure to raise here the standard objection that astronomers do, in fact, witness events which took place a very long time indeed before the actual moment of witnessing, e.g. stellar explosions, which we now witness, took place as long ago as it has taken light to reach us, and we can calculate the time that has lapsed between the event and our witnessing of it. It is, moreover, natural for us to speak of witnessing terrestrial explosions. But, in fact, it is just as natural to speak of witnessing terrestial explosions, even though we know that *some* time, if not as *much* time as in the case of stellar explosions, must have lapsed between the time the explosion occurred, and the time at which we witness it. But we can go further even than this. Epistemologists never weary of pointing out that any perception, of anything, must, for purely physical reasons, occur at some time, however small, after the perceived event itself took place, that it takes *some* time, however little, for an impulse to reach the centres of perception, whatever these may be.[1] But then the example of the exploding star loses some of its force if these facts hold. For the difference only is a matter of degree, such explosions being 'more past' or at a greater temporal distance from a bit of perceiving than ordinary, terrestrial explosions, and these again are only 'more past' than, say, perceiving the bursting into flame of an ordinary match held in our hands. And the question is no longer whether it is possible to perceive past events, but whether, instead, it is possible to perceive anything *but* past events.

Let us try to accommodate these facts as best we can. Today, then, I am able, let us suppose, in the required natural sense, to witness the explosion of a star which is located very many years back in time. In just the same spatial position at which I witness this (e.g. my observatory, affixed to the Earth) I shall be unable to witness this same event *tomorrow*. If I did not witness it today, I should, at this spatial position, never again be able to witness it. Perhaps at a different spatial position I should be able to witness it tomorrow, just as, at a different spatial position I should have been able to witness it yesterday. But in fact I am where I am, and was not,

41

and will not be, in the required spatial positions for having perceived it yesterday, or for perceiving it tomorrow. So this suggests that there is a spatio-temporal range within which an event may be perceived. *E* is perceivable at different times within that range from different positions in that range, and perceiving *E* is a matter of being at the right place at the right time. I can perceive it at different times, but then only from different positions. Now, to say that we cannot witness *E because E* is past is to say (*a*) that *E* took place, and (*b*) that the time at which *E* could be witnessed at the spatial point we occupy is earlier than the present moment. And to say that we cannot now ever witness *E* is to say (*a*) and (*b*) and (*c*) that we cannot reach, at any future time, a different spatial point in time to witness *E*—that is, by the time we reach any spatial point different from the one we now occupy, the time at which *E* could be witnessed from that point would be earlier than the time of our occupying it. One can be within the temporal range, but outside the spatial range for witnessing an event, or within the spatial range but outside the temporal range for witnessing it. Thus someone standing in Strasbourg in A.D. 1066 would exemplify the former case, and someone standing at Hastings in 1963 the latter case. And it is the latter sort of case that is applicable here. It is pointless now for anyone to move into the spatial range for witnessing the Battle of Hastings, for we are forever beyond the temporal range for having the required experience.

With some such emendation, I suppose, we can accommodate the fact that we only witness past events. We can even accommodate the fact that we might at the *same* time witness events which took place at different times, e.g. an astronomer might simultaneously witness a bomb bursting in the air *and* a stellar explosion. We could not tell them apart, of course, by any 'marks of pastness', for if *everything* we witness is a past event, everything must bear marks of pastness, and we should be obliged, on Lewis's suggestion, to speak of one thing as more charged with such marks than another. But little is to be gained by that sort of evasion, for from the fact, if it is a fact, that all we witness are past events, it does not follow that we can now witness *every* past event. For some past events, we are forever outside the required range for witnessing them, and this is now the case with the Battle of Hastings. The question, then, is how we can know unexperienceable past events, events which are really 'dead and gone'. Clearly, it is because we have evidence that they happened. And we may concur that it is on the basis of something which is presently capable of experience that we are able to know about things that happened, but which we cannot now experience. But it might be said that this is exactly Lewis's thesis. Have I given the account, alternative to his own, which he has challenged critics of his view to give? The answer is that I have not. But the difficulties which arose from Lewis's analysis did *not* arise, surely, from the platitudinous claim that we only know about the unwitnessable past on the basis of evidence. It arose, rather, from the claim that when I make a statement about the past I am implicitly predicting the experiences I will have in the future when and if I perform certain actions. I may indeed be implicitly making such predictions. But surely that is not all that I am doing when I make a statement about the past. And Lewis's mistake is to suppose that this *is* all that I am doing, that the whole of my cognitive claims are expressed in conditional sentences of the sort we have recognized.

Just think, for a moment, of how plainly we understand the sentence 'The Battle of Hastings took place in A.D. 1066', and how vivid an image many of us have of this battle. But try now to think of what we could possibly be predicting about our future actions and experiences when we make such a statement. I know very little about the presently available evidence for the truth of that sentence, I have no idea what sorts of things a specialist in English history would produce in support of it. The best I could be predicting, I suppose, is that if I were to ask an English historian for evidence for the Battle of Hastings, he would produce some: but *what* he would produce I can hardly say. If *all* I was doing when I made the statement, was predicting the outcome of such queries, I should have a very unclear idea indeed of what I was saying. And it would be much the same if I were to have said 'The Battle of Waterloo took place in 1815'. I should hardly be able to distinguish between these statements, since I have no better idea of the sorts of evidence I should find in the one case than I have in the other. So even if, in order to be said to know that such a past event happened, I must be able to produce some sort of evidence, the fact is that when I say that such an event happened, I am not merely predicting what my experiences will be as a consequence of seeking for evidence. I am, rather, saying that such and such an event happened.

These are quite different things. My statement was *about* the Battle of Hastings, and not about what one might find in the Royal Archives. I have no idea what one might find there, but at best I would want to say that what one finds there would perhaps be evidence for a statement about the Battle of Hastings, that it would, speaking optimistically, verify it. But if I could not independently and separately speak of past events, what would such verificatory experiences *verify*? Presumably, they enable us to know about the Battle of Hastings. But surely, knowing about the Battle of Hastings is quite a distinct thing from knowing about the evidence for it. To know about the evidence, for instance, might be to experience certain sheets of parchment. But I am certainly not referring to sheets of parchment when I refer to the Battle of Hastings. I am, instead, referring to a scene of human strife. Yet if I were only making predictions when I make my statement, I would be speaking, not of armed men, nor of kings and captains, but instead to scraps of parchment and worn tapestries. And this is an extraordinarily unplausible view. And how, indeed, should I so much as regard these things as evidence for the Battle of Hastings if each utterance of a sentence about the Battle of Hastings turned out to be but a prediction concerning my experience with parchment and tapestry?

Even, then, if the whole of our knowledge of the Battle of Hastings is in some sense based upon such conditional sentences, it cannot *consist* of such statements alone. Lewis is crudely correct when he says that we know about the past because we have evidence for it, and that we know it in no other way. But he has not allowed us a way to talk about the past, but only to talk about that upon which our knowledge of it is based. He has not allowed us a way of speaking about the past which does not immediately become a way of speaking about the present and the future. This is because he is not merely in the grip of the dogma that all we know is what we can experience (and so we cannot know the past), a dogma which has compelled him to introduce all sorts of wildly unplausible entities and marks, but, more importantly, because he was in the grip of a certain theory of *meaning* in accordance with which the meaning of a non-analytical sentence is taken *to be* the set of experiences which verify it. It is to this, accordingly, that I must turn.

'For my own part', Ayer wrote, in the heyday of verificationism, 'I do not find anything excessively paradoxical in the view that propositions about the past are rules for the prediction of those "historical" experiences which are commonly said to verify them, and I do not see how else our "knowledge of the past" is to be analysed.'[1] He adds that he suspects those who feel dissatisfied with such an analysis to be tainted with the metaphysical view that the past is 'somehow "objectively there"'—that 'it is "real" in the metaphysical sense of the term'.[2] Though it is worth pointing out that it is precisely such a metaphysical assumption which seems to have tormented Lewis, namely, that since the past is *not* 'objectively there,' it cannot be experienced, and hence cannot be known, or that, at all events, we can only know what is 'objectively there' and so what we know of the past must be knowledge of something 'objectively there', and hence not the past. Nor can Ayer himself be very far from holding to such a view, namely, that there must *be* something for our statements to be about, which we can experience, if we are to know them, and so, if we are to know sentences about the past, they cannot *really* be about the past, but about something we can experience. Despite his brave words, Ayer retreated from the position that statements about the past are not statements about the past, but, rather, rules for making statements about the future.

One reason why a reasonable man might wish to retreat from this position, his taste for paradox notwithstanding, is that it entails a revision of what a sentence like 'The Battle of Hastings took place in A.D. 1066' *means* each time that sentence is verified.[3] That is to say, most of us might allow that in *some sense* 'The Battle of Hastings will take place in 1066' differs in meaning from 'The Battle of Hastings did take place in 1066'. Perhaps we would say this because the former could have been, while the latter cannot be, verified by experiencing the Battle of Hastings (though in fact few of us would give this as a reason for saying they differ in meaning). But who would wish to say that 'The Battle of Hastings took place in 1066' differs in meaning from 'The Battle of Hastings took place in 1066'? Yet this is what we might be committed to say on the original verificationist analysis: the sentence changes its meaning each time it is verified. For suppose it is taken at one time as a prediction that a certain experience will be had, and this experience *is* had. Then it can no longer predict *that* experience, but another one and so its meaning changes. We can give in to our prejudice that it has always the same meaning, only by the artificial means of using it to predict an experience to take place after the absolutely final utterance of the sentence. But in many cases it is too late for that. Thus 'Caesar died' no longer means what it once did, partly as a consequence of the meddling inquests of Marc Anthony. So the theory induces a radical instability in the meaning of most sentences about the past, or at least all of them which ever have been verified. Indeed, in a kind of Heraclitian way, we could never verify the same sentence twice. We should always, instead, be verifying a different sentence, in case difference of meaning means difference of sentence. And this would entail that 'Caesar died' and 'Caesar died' are not the same sentence in the case where one of them has been verified. Yet we surely want to say that these are both statements of the same sentence, and that this same sentence has always the same meaning. Nor would it help much to say that these are different *uses* of the same sentence to make different *statements*. For these different statements could never mean the same if one of them were ever in fact verified, or if they were verified by different experiences.

Ayer came to admit that it is misleading to suppose that statements about the past 'can be translated into propositions about present and future experiences'.[1] He said that 'this is certainly incorrect', and added that he did not any longer think that 'the truth of any observation-statements which refer to the present or the future is a necessary condition of the truth of any statement about the past'. But the question had not to do with truth, but with meaning, and he remained concerned with how such statements could be regarded as meaningful if we could not, by experiencing what they were about, verify them directly. He answered this by introducing the notion of 'verifiable in principle'. This involved a shift in programme. Sentences about the past were not to be translated into sentences about present and future, but they were to be translated from the indicative to the subjunctive mood. And I want to examine this notion now.

It is true that I, who have occupied a stretch of time beginning in 1924, and continuously since then without interruptions in my existence, can never have observed events which happened before 1924, or whose spatio-temporal ranges fall short of that date at their forward edge. But during that time, I have occupied a variety of different spatial positions. While in those spatial positions at a given time, I could not have observed events contemporary with my occupying them if, at those times, I was outside their *spatial* range. In Rome in 1962, I was unable to witness things going on in New York. But I could have been in New York in 1962, rather than in Rome. It is not logically absurd to suppose this. And had I been in New York, I might have witnessed what happened there. It was a sheer contingent fact that I was one place rather than another. But it is exactly the same with time. I might have lived in a different stretch of time than I, as a matter of sheer contingent fact, have occupied. And just as it is not absurd to suppose that instead of having been in Rome in 1962, I might have been in New York, so it is not absurd to suppose that instead of having been in 1962 in Rome, I might have been there in 44 B.C. Just as it is a contingent fact that I did not witness New-York-events in 1962, but Rome-events instead, so it is a contingent fact that I did not witness 44 B.C. events in Rome, but 1962 events instead. I did not witness those events, but it is not logically absurd to suppose that I might have. So I cannot in fact verify Caesar's death by witnessing it. But I could have verified it had I been there at the time. So the sentence 'Caesar died' is verifiable *in principle*. And, since it is verifiable, it is meaningful. This, roughly, is Ayer's analysis here.[1]

Let us waive the question whether I would have been the same person I in fact am if I had been in 44 B.C. instead of A.D. 1962. Let us only ask whether this new account manages to avoid the dizzying shifts in meaning upon which its predecessor foundered. In a way it does avoid that. We are to take all instances of 'Caesar died' as having reference to the same set of possible experiences, namely those one would have had, had one been in Rome in 44 B.C. Once more, there is a *translation* thesis here, but instead of a translation into a conjunction of conditional sentences,

(1) Caesar died in Rome in 44 B.C.

is to be translated into some such sentence as,

(2) If I had been in Rome in 44 B.C., then I would have had Caesar-dying experiences.

47

Now (2), as we shall see, is not a full or perfect translation of (1), but it will do well enough for present purposes. Notice, I am no longer referring to my present and future experiences when I want to refer to the past. Indeed, I am no longer obliged to be referring to any experiences I shall ever, in actual fact, have. On the other hand, I am not quite able yet to refer to Caesar's death. Instead, I am obliged to refer to experiences I would have had, had I been in a certain place and at a certain time. We must, of course, not be put off by the possible objection that no two speakers of (1) can mean the same thing since (1) is used, in each case, to speak of the speaker's *own* subjective experiences. This can be avoided I think, quite easily, by suggesting that the experiences in question would have been had by *anyone*, that had you been there instead of me, you would have had these experiences, so that these experiences may indifferently be referred to by each user of (1). We may now consider, as a better (partial) translation of (1)

(3) If anyone were in the appropriate place (etc.), he would have had Caesar-dying experiences.

'Caesar-dying experiences' is perhaps a bit makeshift. It stands roughly for the experiences which would directly verify the sentence 'Caesar dies now'. I remark, parenthetically, that (3) is therefore not strictly speaking true, for it fails to be true for Caesar himself: 'Death', as Wittgenstein wrote, 'is not an event in life. Death is not lived through.'[1] But I shall not press this point, for I am still concerned that we cannot yet talk about Caesar's death—but only about Caesar-dying *experiences*. The reason this *is* a makeshift term is that we don't have, in our language, words which will do the precise job which is required of them by the analysis being considered. It is, rather, to be the job of a different language altogether than the one we speak, a language in which all terms which refer ordinarily to physical events and objects are to be rendered into other terms which refer to experiences. This explains why (2) and (3) are only *partial* translations: 'Rome' designates a specific physical city, and a complete translation would replace 'Rome' with whatever experiential equivalent it is to be provided with in this new language. We are dealing, in other words, with the programme of Phenomenalism. And this is why we have found it hard to refer to Caesar's *death*—a *physical* event. The acknowledged collapse of verificationism in its original form does not, Ayer writes,

Mean . . . that propositions referring to the past cannot be analysed in phenomenal terms; for they can be taken as implying that certain observations would have occurred if certain conditions had been fulfilled. But the trouble is that these conditions never can be fulfilled; for they require of the observer that he should occupy a temporal position that *ex hypothesi* he does not.[1]

But, as we have seen, the latter difficulty is not insuperable. Nevertheless, we had better look for a moment at what is really involved in the proposed translation.

Phenomenalism is the thesis that all statements purportedly about physical objects and events must be translatable into sets of statements about actual and possible experiences if they are to be meaningful. The pretence is that I can only understand a term if I know what experiences I would have if I were confronted with the designatum of the term. But then the term must be rendered intelligible by other terms which refer just to these experiences, and nothing can meaningfully be said of the designatum which cannot be so rendered. A full-scale discussion of this programme is not appropriate here, but from this bare statement of it one should be able to see why we were unable to refer to *past* events, in this case the death of Caesar. This is a consequence of the fact that, according to the phenomenalist, we cannot refer to events *simpliciter* if, by events, we mean physical occurrences. For any attempt to refer to an event immediately involves us in reference to actual or possible experiences. So it is not some special fact concerning the past which prohibits reference to past events. It is, rather, a general fact that we cannot refer to events, as physical occurrences, and hence, as a trivial consequence, cannot refer to *past* events. Even Brutus would have been unable to refer to the death of Caesar, but only to 'Caesar-dying experiences'. So this is not a special problem which arises for statements purportedly about the *past*.

I shall not tarry over the question as to whether we can or cannot, in fact or in principle, effect a phenomenalistic translation, a translation into terms referring only to sense-data and to sensibilia—to actual and possible experiences—a statement which refers ostensibly to the assassination and death of Caesar. I do not know whether it can be done, but I shall assume it can be, and that we have succeeded in doing so. But I am not sure I can

understand how the phenomenalist would render *pastness* in phenomenalistic terminology. Ayer has suggested a way in which we can speak
of it as *possible* that we might have had the experiences referred to in a
phenomenalistic rendering of the expression 'The death of Caesar'. It is
possible in the sense that it is not logically impossible that we should have
been in Rome in 44 B.C. But he has said that *in fact* the conditions for
having these experiences cannot be fulfilled, presumably because we
cannot in fact occupy the required time-space position: 'they require of
the observer that he should occupy a temporal position that *ex hypothesi*
he does not.' Yet one cannot but point out that reference to temporal
positions is surely reference to a physical location, and that until we are
shown how the notion of temporal and spatial positions are to rendered
in phenomenalistic terms, we must assume that at least some meaningful
physical notions do not have experiential equivalents, and if some things
cannot be put into the favoured idiom, we have no good reason for
accepting the cumbersome locutions of phenomenalists at any point. A
partial phenomenalism is philosophically worthless, for the *claim* is that
whatever is meaningful in discourse is meaningful in terms of actual and
possible experience. Comparably, there are *some* angles which we can
trisect with ruler and compass. But this does not establish the general case,
even though a proof that there is any angle incapable of such trisection
*dis*establishes the general case. And if we cannot capture temporal positions by means of phenomenalistic predicates, this means the complete
collapse of phenomenalism.

But I continue with my supposition that we have achieved our translation of the death of Caesar, and I shall even assume we have taken care of
putting into phenomenalistic language reference to time and space
positions. So the sentence

(4)   Caesar dies in Rome in 44 B.C.

is successfully rendered by the sentence

(5)   If anyone were to have Rome-in-44-B.C. experiences, then he would
have Caesar-dying experiences.

It makes little difference that (5) is sketchier than a full translation would
be. I shall just suppose that it marks the place and sets the form for the fully

adequate translation, however long and complex the latter may be. For the issue which concerns us really lies elsewhere. It lies, indeed, precisely here: one cannot tell *when* the sentence (4) was uttered, nor whether it refers to something past, present, or future (forgetting that Romans would not have employed the expression 'B.C.'). Nor can one tell this from (5). And the reason for this is because (4), as (5) shows, has been put in a tenseless idiom. What I am interested in is how we put into experiential terms the fact that a given event is *past*. And this, we shall see, is quite a different question from the one that asks how we can put into such terms reference to time-space positions. For we might have succeeded in the latter task without being able to tell whether the time-space position thus translated is past, present, or future.

It is sometimes objected against Phenomenalism that a sentence such as (4) could be false, even though a sentence like (5) were true. Thus there may be no daggers before me though I have dagger-experiences. This objection has little force, however, if the phenomenalist is correct in maintaining that (5) says nothing which (4) does not, that it is only a translation of what is meaningful in (4). Nevertheless, we are entitled here to a comparable and, I hope, a more telling criticism. Notice that (4) is tenseless, and (1) is not. But then it would be a mistake to regard (5) as indifferently a rendering both of (1) and (4). Because, since (1) contains information which (4) does not, (5) is either an inadequate rendering of (1) if it is an adequate rendering of (4), or it goes beyond a translation of (4) if it is an adequate translation of (1). And (1) *does* give information that (4) does not, specifically that the event referred to by each of them took place in the *past*. (4) does not tell us whether the event has happened, is happening, or will happen. Hence (1) could be false while (4) is true: it could be false if the event had *not* happened in the past. So, if (5) is supposed to be an exact translation of (4), (1) could be false while (5) was true, since (5), no more than (4), contains all that (1) does. But more generally, it is possible for *any* tensed sentence to be false even though its phenomenalist translation is true. Unless, of course, we could render tenses phenomenalistically.

It is not easy to see how we should render tenses in experiential terms. One might of course, propose some such stratagem as this: giving an experiential equivalent for *moving through time*.[1] Thus:

we arrive at 44 B.C. by traversing a series of event-stages, and each of these can be rendered phenomenalistically. True, we cannot occupy these positions, but it is possible that we might have, for the reasons already considered. The difficulty, however, would be in making the *first* step between here and 44 B.C. For the first step must be to an event which is past if our trip is to be in the right direction, and the question is how we indicate that the first step is in the direction of the *past*, or differs from a first step in a *future* time-journey. One could say: the first step in the direction of 44 B.C. But then we have to express somehow that 44 B.C. is in the *past*, and this then begs the question. Certainly you cannot hope to do the trick in a *tenseless* idiom. For suppose we say that 44 B.C. is 2007 years before *now*. But 'now' indicates the use of a present tense, and would have to be replaced with a date, that is A.D. 1963. We might then say that the statement that 44 B.C. is 2007 years before A.D. 1963 is true, and, for that matter, analytically true. But this does not tell us that 44 B.C. is past. For someone could have uttered this truism at any time, including 43 B.C., when the years referred to were future. We have to know when the sentence is uttered, and then whether this time is before or after or concurrent with the time at which we raise the question. So we cannot readily eliminate the sort of information tense gives us. But then, if we cannot incorporate this information into our phenomenalistic translations, phenomenalism collapses as a programme for expressing all that is meaningful in our ordinary language. To be sure, one could take the heroic course of saying that tense-information is meaningless, but this is unreasonable, for surely we understand what is meant by saying that something is past. At this point a phenomenalist will challenge us, as did the epistemologist in an earlier discussion, to say how, if *not* in experiential terms, we do understand this information. But I shall not even try to say this for the present. I will return to it in a later discussion, and take it on faith that the expression 'for the present' and 'later' will be understood, however they are to be analysed.

Phenomenalism, as much as verificationism, involves an attack upon the success of the minimal historical aim, for we cannot, if the former is correct, make a meaningful statement about the past which does not immediately become a statement about actual and possible experiences.

But I have used this very point to mount an attack upon Phenomenalism itself. If it cannot succeed in accommodating the sort of information we get in tensed sentences into the idiom it favours, this would be a defeat for Phenomenalism. But the exact status of tenses remains to be clarified, and I shall make a first step in this direction by engaging in yet one more polemic; one which will bring this entire discussion to a head. Professor Ayer, whose efforts at analysing sentences about the past have been tireless, has recently produced a distinctive analysis which justifies, if it is correct, the claim that, in his words, 'No statement as such is about the past'.[1] This claim, of course, would once more spell defeat for the minimal historical aim, and it would rescue Phenomenalism from the straits we have cast it into. For if no statement is *as such* a statement about the past, it can hardly be a defect in Phenomenalism that it cannot translate into its own idiom statements about the past. One cannot translate non-existent sentences. But then our minimal characterization cannot be fulfilled either, since there is no sentence of the required sort for historians to succeed in making. Nevertheless, one wants to know what, if not the past, a statement like 'Caesar died' is about. It is just this that Ayer's new analysis seeks to answer.

To begin with, Ayer allows that when we use this sentence, we are referring to an event, in this case, the death of Caesar. We are not, however, referring to a *past* event, for events, as such, are neither past, present, or future. So, 'considering only the factual content of state-ments',[2] when we refer to Caesar's death, we are referring to an event, but not to a *past* event, for, apparently, the expression 'past event' in some sense involves a mistake of category. This sounds, perhaps, more puzzling than it need sound. It amounts to little more than insisting upon a distinction between one-place and multi-place predicates, or between absolute and relational properties. If it sounds paradoxical, it does so in exactly the same way that the following claim does: that no statement, as such, is ever about anything which is next to something else. True, a bottle may be next to a box, and the statement which asserts this would be true. But in the sense in which it is correct to say that a bottle is green, it makes no sense to say that a bottle is *next to*. Bottles, as such, are not *next to* or *between* or *behind*. So a statement to the effect that a bottle is next to a box is a statement about a bottle, but *not* about a 'next-to' bottle. For there are no such things. Similarly, then, a tensed statement is about an event, but not a *past* event. For being past is not a property of events, but a relationship in which events may stand as one term. The *factual content* of such sentences has reference to events and to absolute properties of events. If we subtract, from a tensed sentence, this factual content, we are left something which, strictly speaking, indicates the temporal position of the person who utters the sentence vis-à-vis the event to which the sentence refers. We do not have, as grammatical features of our language, in the way tenses (in this analysis) indicate the temporal relationship in which we stand to the events we refer to, devices which automatically indicate the spatial relations in which we stand to the things we refer to.[1] But when I say that the door is on my left, 'on my left' is not a property of the door I refer to, but a relationship between the door and myself. Someone else might say, of that same door, that it is to his right. But the two statements 'The door is on my left' and 'The door is on my right' are not inconsistent, even if asserted of the same door and even if asserted at the same time, providing they are asserted by different persons standing in different spatial relations to the door. But even different persons, supposing them to be speaking of the same door at the same time, *would* be asserting inconsistent statements if they respectively said 'The door is wooden' and 'The door is metal'. But similarly, there would be an inconsistency between 'Caesar died in 44 B.C.' and 'Caesar was alive all through 44 B.C.' if asserted by different persons at any time. But 'Caesar will die in 44 B.C.' and 'Caesar died in 44 B.C.' are by no means inconsistent if uttered at different times by the same or by different persons. Indeed, one feels at first glance, if one of them is true, the other must be true, and if one of them is false, the other must be false, so that, by definition, far from being inconsistent, they are materially equivalent.

It follows, then, that tensed sentences may be analysed into two distinct components, each of which gives a different piece of information; one having to do with an event, and the other a relationship between an event and a time at which the statement is uttered. The following three sentences, accordingly, uttered respectively by Calpurnia, Brutus, and Marc Anthony, are indifferently about the same event: (*a*) Caesar will die; (*b*) Caesar dies now; and (*c*) Caesar died. Concerning just the *factual* content of these three statements, it makes no difference when they are uttered, for the tense has no bearing on the factual part of the sentence. The two pieces of information conflated together in a tensed statement are 'logically distinct', and the three statements are equivalent: if one is true, all are, and if one is false, all are.

This, if I understand it correctly, is Ayer's analysis. I wish to argue that notwithstanding its ingenuity, this analysis is not wholly sound. Those three sentences are not equivalent, and the component pieces of information brought together in tensed sentences are not logically distinct if, by this, we mean 'logically independent'. My argument, if sound, will entail that one cannot quite so neatly extricate the temporal from the 'factual' information contained in tensed sentences.

Let us begin by considering the following claim, which states, succinctly, the thesis I have been describing:

The truth or falsehood of a statement which purports to describe the condition of the weather at a given date is quite independent of the time at which it is expressed. By combining a description of the event in question with a reference to the temporal position of the speaker, the use of tenses brings together two pieces of information which are logically distinct. It does this in an economical fashion, but it is not indispensable. Either piece of information could perfectly well be given in a language that contained no tenses at all. The temporal position of the speaker, relatively to the event described, which is shown by this use of the present, past, or future tense, could itself be characterized by being explicitly assigned a date.[1]

This seems to me to license the view that a tensed indicative sentence is analysable as a truth-functional conjunction of logically distinct propositions, the conjunction being obscured by sheer grammatical accident. One conjunct (A) says something about an event E, and the other conjunct (B) says something about the temporal position of the speaker relative to E. Either piece of information could be given separately, and since we are supposing the two conjuncts to be logically independent, the truth or falsity of either of them leaves undetermined the truth-value of the other. Of course, the truth or falsity of the conjuncts will have something to do with the truth or falsity of the conjunction taken as a whole: this follows from our supposition that a tensed indicative is a disguised *truth-functional* conjunction. In particular, the conjunction will be false if either or both conjuncts are false. In this case, it naturally follows that the truth value of a tensed indicative will *very much depend* upon the time at which it is uttered, for this will be one of its truth-conditions. For example, if Brutus utters (*b*), his statement will be false, if Caesar had already expired, or if he has not yet done so. Brutus's statement, we are supposing, would have been in the present tense. But then his statement will be false because one of its conjuncts is false: in this instance the conjunct which refers to the temporal position of Brutus, at the time of utterance, relative to the event he describes. He will have mis-stated the relationship, saying, in effect, that the utterance is concurrent with the event it is about, when in fact it is later or earlier than that event. In this regard, then, it is plain that the three sentences are not equivalent: (*b*) can be false, though (*a*) is true or (*c*) is true. Thus these two sentences do contradict one another if we take into account the time of utterance: (*I*) 'Caesar will die.' (*II*) 'No, he has already died.'—even though each has the same 'factual content.' If (*a*), (*b*), and (*c*) are uttered at the same time, two of them will be false if one of them is true.

In the case considered then, the conjunct (*B*) is false, and the conjunction *as a whole* is, accordingly, false as well. But of course it may be said that this still leaves undetermined the truth-value of the other conjunct (*A*), understood tenselessly about Caesar's death. One might then say that (*A*), if true, is true independently of the time of its utterance, and that (*A*) is therefore independant of the other conjunct (*B*). This is doubtless what Ayer had in mind: the truth of a tenseless proposition does not depend upon the time of its utterance. And doubtless he was thinking of untensed sentences when he said that no sentence, as such, was about the past. But a tensed sentence very much depends, for its truth-value, upon the time of its utterance. It follows, then, either that we cannot give a tenseless rendering of tensed sentences, or that some tenseless sentences very much depend, for their truth value, upon the time of their utterance. So one or the other part of Ayer's analysis has to be rejected. But it is very hard to suppose that a sentence which is about the time of its own utterance does not depend upon the time of its utterance. It would be exceedingly awkward to suppose that 'This sentence is uttered at *t*-1' does not have its truth value determined by the time at which it is said. It can hardly be timelessly true. Even, then, if we introduce the 'explicit date' into the sentence itself, we have not succeeded in both rendering it tenseless *and* independant of the time of its utterance.

Notice, moreover, that '. . . is independent of . . .' is not a symmetrical relation. Even if the conjunct (*A*) is independent of the conjunct (*B*), the converse does not follow. It may or may not be independent, but in fact it can be proved that it is not independent. And if this is so, it cannot be the case that we can give one piece of information independently of the other, as the truth-functional interpretation suggests. For let us suppose that (*A*) is *false*. 'Caesar dies in Rome in 44 B.C.' could be false in a number of ways: if there were no such person as Caesar, if Caesar were immortal, if Caesar dies at some other time or some other place. In any case, to suppose the sentence false is to suppose that there is (tenselessly) no such event as the statement purports to describe. Now if (*A*) is false, the conjunction, of course, is false. But the question remains how the other conjunct (*B*) could be true if (*A*) is false? How can I stand in *any* temporal relationship with a non-existent event? The relation collapses for want of a term. One might, of course, say that it is a *fact* that Caesar does not die in 44 B.C. But 'facts themselves are dateless', and I cannot then regard a statement as being uttered before, or after, or concurrently with something to which no date can sensibly be assigned. The truth of (*A*) is thus a necessary condition for the truth (or, on an alternative analysis, for the truth or falsity) of (*B*). One might then say, if one wishes to, that the truth of a tensed sentence presupposes the truth of that part of the sentence which may be stated in an untensed way. Nevertheless, a tensed sentence may be false when the untensed component is true, and this shows that they are not equivalent. But moreover, we find here just the same sort of situation we discovered in connection with Phenomenalism: a phenomenalist rendering of that part of a sentence which can be rendered tenselessly may be true while the corresponding tensed sentence is false. And the information which these give us cannot be phenomenalistically rendered. In so far as we are unable to eliminate tenses in such a way that this same information may be stated tenselessly, we are hardly entitled to the view that no statement is, as such, about the past. A true sentence in a past tense is, as such, about the past.

Notice, finally, that this same situation arises if we think of rendering, in a somewhat different way, the two pieces of information conveyed in a tensed sentence. The only natural alternative to the truth-functional conjunction which I can think of is this. We may regard tenses as operators, specifically, as statement-forming operators which makes statements out of statements. As operators, they of course have no truth-value standing on their own, for example, in the way in which the quantifying operator (x) is not, as such, either true or false. Now let p be an untensed sentence, and let P be a tense-operator which has the force of putting p in the past tense. Thus $P(p)$ says: 'It was the case that p.' Now it could be the case that p is true, and $P(p)$ is true; or the p is false and $P(p)$ is false. What cannot be the case is that p should be false and that $P(p)$ should be true.[1] Nor, more generally, can it be the case that p is false and $T(p)$ true for any value of T, if T be considered an operator-variable which takes tenses as values.

A good many of the problems I have been concerned with, of course, really arise out of the concept of truth. It is not so much that 'Caesar dies in 44 B.C.' is tenseless, but that 'It is true that Caesar died in 44 B.C.' is taken as tenseless—largely because truth is regarded as an atemporal fact regarding sentences. Hence if 'Caesar died in 44 B.C.' is true, then it must be *timelessly* true. So regarded, the time at which it is uttered is apparently irrelevant: if timelessly true, it would be true whether uttered before, or during, or after 44 B.C. And this then renders tense somehow otiose. The idea that truth is atemporal is, however, a singularly mischievous notion, and I shall later[2] seek to give reasons for rejecting it. But for now I wish only to say a few words more concerning the analysis I have just examined.

Why should Ayer wish to say that no sentence as such is about the past (or, for that matter, about the present or the future)? I suggest that the refusal to take tenses seriously is due to the fact that Ayer remains haunted by the old question of the verifiability of sentences about the past. His strategy consists in trying to show that this problem need not arise, inasmuch as no sentence *is* about the past. And so there is no real problem concerning the verifiability of historical sentences. These sentences, not, on his analysis, *being* about the past, are therefore not threatened by the objection that they are not verifiable because what they are about is *past*. Ayer's claim is that they are about events, but not about past events. The truth or falsity of such statements depends then wholly upon what is (timelessly) the case with the events they are about, and not upon the time at which they are uttered. 'A sentence', he writes, 'which is verifiable when the event to which it refers is present is equally verifiable when the event to which it refers is past or future.'[1] But what he means to say is that an untensed sentence, if it is ever verifiable, is always verifiable, that its verifiability is not a function of the time at which it is uttered. True, this formulation does suggest that the sentence's timeless verifiability depends upon its being verifiable at some *time*, and that, unless the sentence in the present tense is *sometime* verifiable, namely at the time the event referred to occurs, it is not verifiable ever. But this is not the point I wish to insist upon. I wish rather to emphasize that this is so for untensed sentences, but does not establish that *tensed* sentences are verifiable. The truth of an untensed sentence does not guarantee the truth of all tensed versions of it. And it may very well be that the *verifiability* of an untensed sentence does not guarantee the verifiability of all tensed versions of it. After all, the allegedly verifiable content of these sentences is only a *part* of the whole sentence, by Ayer's own analysis. And the verifiability of a part does not entail the verifiability of the whole: the verifiability of 'grass is green' does not entail the verifiability of 'The grass is green and the Tao is purple'. There is *still* room for scepticism about the past.

But in fact such scepticism is quite independent of the entire issue Ayer has spent so much time and effort and ingenuity on in the hope of defeating it: as though he had mounted an army on the wrong battlefield. That the verifiability of untensed sentences has nothing to do with the time at which they are uttered is just what we would expect, given that verifiability is a matter of *meaning*. We are to understand a sentence with reference to the sorts of experiences which would be required for it to be verified. However, one may concur that whether a sentence is meaningful does not depend upon the time of its utterance. A sentence may be meaningful even if, in fact, there is nowhere in the timestream an event or entity for it to refer to. If there is one point most contemporary philosophers are persuaded of, it is that there is a difference between the meaning and the reference of a term. If it should prove true that there was never such a person as Caesar, sentences purportedly about Caesar would not collapse into meaninglessness. False sentences are not meaningless, nor are fictional sentences: we understand *Hamlet* as readily as we do *Julius Caesar*. I may say that the birth of Caesar's seventeenth daughter by his ninty-sixth wife is celebrated sesquicentennially by the hooved brewmasters of Lebanon, and this sentence, false if any sentence is, is nevertheless meaningful and, for that matter, even verifiable. Surely the predicate 'is verifiable' is not to be restricted to *true* sentences alone. It would be peculiarly self-defeating if being true were a necessary condition for being meaningful, for how should we know then whether a sentence were meaningful unless we first ascertained it were true? But then how should we ascertain it were true unless we knew what it meant? For sentences in the past tense, we could only say that they were meaningful, or verifiable, if we first knew that what they were about actually took place. The attributing of meaningfulness to such sentences then would presuppose knowledge of the past.

Meaningfulness, understood as verifiability, is independent of the truth-value, of the referring relations, and of the time of utterance of sentences. But if, by meaningfulness, we mean verifiability, the question remains how we are to understand the meaning of *tensed* sentences. What experiences verify that what we are speaking of is *past*? This is the problem one found in Lewis, and in the Pragmatists generally, and in Phenomenalism, the problem of defining, in experiential terms, the sort of information furnished solely by the tensed parts of sentences, once we have subtracted the 'factual content'. And at this point one cannot but feel the attractiveness of the Kantian position, that time is not a datum of experience, but a form of experience, a precondition for experience. And the frustration we have continually encountered is reminiscent of the celebrated difficulties Wittgenstein made so much of in the *Tractatus:* how are we to put into language the relationship between language and what it is about? If 'aboutness' is a relation, we can put it into language only by putting its terms into language, and this destroys the relationship between language and the *world*. Reference is not part of language, but parts of language constitute one of the terms of the referring relationship. Comparably, the assertion of a sentence is not *part* of the sentence asserted. Pragmatism, and Phenomenalism as well, are attempts to suck the whole of reality up into experience, or into language. What we have experienced as a continuing frustration is in fact a limit to such a programme. In this regard, oddly enough, one might go so far as to say that after all, tenses are not parts of the sentences we assert. One might regard them, instead, as ways of asserting that a certain sentence is, or was, or will be true. And this would be very like saying that the truth of a sentence is not part of the sentence. But the difficulty here is that tenses reappear in the expressions 'is true', 'was true', and 'will be true'. So an analysis of them remains to be given: an existentialist would say that they register the way in which we are in the world of time.

There remains one point more. In this discussion, I have acquiesced in the identification of meaningfulness with verifiability, and have indicated that perhaps part of our understanding of a sentence has to do with our knowing what experiences would verify the sentence. Ayer has said that if a sentence about an event can in principle be verified at the time the event occurs, it is forever verifiable. But this suggests that sentences about the past must be of a kind which could be verified by a witness to the event in question. And this, I am afraid, is too much of a concession. For many, and perhaps the most important kinds of sentences which occur in historical writings give descriptions of events under which those events could not have been witnessed. Petrarch's brother witnessed Petrarch's ascent of Mt Ventoux. Historians might say that when he climbed Mt Ventoux, he opened the Renaissance. But his brother could not have witnessed Petrarch opening the Renaissance. He could hardly have seen the event under that description, not because his senses were defective, but because he could not have understood the description at the time. Not unless he knew what was going to happen in the future, and knew, in addition, what historians were later going to say was the significance of what he saw. What experiences would verify for him, at that time, the sentence 'Petrarch is opening the Renaissance'? I would hardly dare say. I should like to say that however meaningful such a sentence is now, in its appropriate past tense, it would have been on the verge of meaninglessness when the event referred to by it was happening. For strictly speaking, there *are* no experiences which verify that sentence, if, by verification here, we mean experiencing what the sentence is about under the description of it given by the sentence. Verifiability, then, is not an adequate criterion of meaningfulness so far as these historical sentences are concerned.

The philosophical importance of these sentences, then, is this. If there are true descriptions of events under which those events cannot be witnessed, our incapacity to witness those events has, with this class of descriptions, no bearing whatsoever. For even if we could witness them, we could not verify them under *these* descriptions. The *general* analysis of sentences about the past has hardly been broached.

V

TEMPORAL LANGUAGE AND TEMPORAL SCEPTICISMS

Should a man choose to be sceptical of sentences purportedly about the past, he would hardly be daunted by the consideration that such sentences are meaningful or verifiable in principle. Their meaningfulness he might grant out of hand, inasmuch as this is a condition for the intelligibility of his own position. Fictional sentences, after all, are meaningful even if false, and the sentences which go to make up an historical novel are of a piece with those which go to make up a proper work of history. What the sceptic is concerned to do is to challenge us to distinguish between the two classes of sentences. Imagine someone mixing up the history books with the historical novels—or with any kind of novel for that matter—and then asking us to sort them out, by criteria, let us suppose, internal to the books themselves or the sentences which go to compose them. The mere label 'history book' will not help, nor the appearance of the word 'history' in the title. A novelist may employ, in a *roman à clef* the familiar disclaimer that all its characters are fictitious, and resemblances between their situations and those of actual persons are pure coincidence. Or a novelist might write 'All that I am about to say is true, so help me God!' And the first book may be true and the latter pure phantasy. Or a man might write out of his wild imagination a sheer confection which he subsequently discovers, to his horror, to be gospel truth. We speak of things coming true. But we can as easily speak of things *having* come true, the events, as it were, happening before the statements describing them have been made, when he who made the statements had no idea that he was speaking truly. It is not fiction being true, however, which concerns the sceptic, but rather history, or what passes for history being false. He is prepared to say that we cannot tell whether it is so or not. We could hardly sort out the shuffled books by the criterion similar to the one appealed to by Hume when he was seeking to sort out memories from images. For novels are on the whole far more vivacious than historical works. Meanwhile, the quasi-aesthetic criterion of relative dullness seems somehow insufficient for certifying the truth of stories.

The fact is, of course, that we cannot, special instances apart (and these are of primary interest to logicians), distinguish true from false sentences merely by inspecting, so to speak, the surface of the sentence. For truth has to do with a relationship between sentences and whatever it is that they are about. The sceptic, thinking in terms of having independent access to what sentences are about, and then seeing, by inspection, whether the sentences are true, will argue that we do not have the required access to what historical sentences are purportedly about to determine whether or not they are true. And so we cannot know. True, we have evidence, and make inferences about the past on the basis of this. But, again thinking in terms of inspection, the sceptic contends that we have no way of finding out finally whether our inferences have connected with fact. And so, once more, we cannot know. Concerning these questions, issues of meaningfulness hardly enter: though Pragmatism, and Phenomenalism in a way, could be read as attempts to circumvent scepticism of this (and other) sorts. If we reject *them*, we must meet the sceptic head on.

To be sure, the sceptic can hardly appeal to the fact that we stand at a certain temporal distance from an event when he makes his strictures. He cannot say, for instance, that we cannot know that E has happened *because E is past*. For we cannot assert that E is past without presupposing the very thing which is apparently to be called into question. If there was one result of our recent discussions, it was that a tensed sentence presupposes the truth of a corresponding untensed sentence. To say that E is past is then already to presuppose the truth of a sentence to the effect that the event E (tenselessly) takes place at a time t, and that t is earlier than now. But if we accept that E has happened, what further can be wanted by the sceptic? We cannot say both that E is past and that we know nothing about E. For we know that E is past. To say as much goes beyond what a scepticism ought to allow itself. To indicate the reasons why we cannot inspect E (because it is past) is to take for granted the truth of at least one sentence about the past, namely that E has already happened and cannot be witnessed. But if we allow ourself that much liberty, then it is plain that some statements about the past can be made even though we cannot witness what they are about. So what is there to insist upon? Considerations such as these suggest that a scepticism about the past which presupposes the very sort of facts it says cannot be established, is a scepticism of negligible philosophical interest. This is especially so in view of the considerations brought out in the final paragraph of the last chapter, namely, that some of the important descriptions we give of past events are such that under those descriptions we could not anyway *inspect* the events they are about.

Scepticism leaves intact the rules of meaning in our language, and attacks instead the rules of reference. It does not say that there are things about which we do not know, but asks, instead, whether there *is* anything for what we say to be about, or whether we have any way of knowing that there is. Scepticism derives its force from the fact that it leaves experience just as it finds it, changing nothing, but only asking whether experience itself relates to anything. And since what it is that experience (or language) is to relate to is not itself *part* of experience (or language), experience (or language) is left untouched. A scepticism concerning the past would have to leave everything just as it is, leave all the techniques for establishing historical statements just as they are, and ask questions which undercut these techniques, which are beyond the reach of these techniques altogether, so far as the answering of these questions is concerned. And since it is only by means of these techniques that we can answer questions about the past, these sceptical questions about the past cannot be answered. This does not mean that scepticism is invulnerable, but it does mean that scepticism about history cannot be settled by history itself. Scepticism nevertheless reveals something about history, if only a limit, and philosophy, concerned with limits, can justifiably examine it.

The argument that, for all we know or can know, the world might have been created, *ex nihilo*, just five minutes ago, raises an initial question for us, the question, namely, of what possible difference it can make to us that there should, in fact, have been anything before that time. For the argument supposes that things would be just as they are, and we should behave just as we do, though the world itself, in which this behaviour is being carried on, is but five minutes old. We should, for instance, have all the memories we in fact have, though most of our memories, *all* of our memories which pretend to be of events which took place more than five minutes ago, would be *false.*[1] The events we seem to remember just never happened. But, since these are our memories, and are taken as such, what difference would it make if in fact they all were false? Again, we would still consider the same persons as our parents whom we now consider to be our parents, though strictly speaking everyone in the world, except a few newly born babes, would be of precisely the same age. Stylistic differences would still exist amongst the artifacts surrounding us, though Carcasson and Delphi would be no older than Levittown, and the *Merode Alterpiece* no older than the *Demoiselles d'Avignon.* Rocks would contain fossils, bronzes would bear the patina of antiquity, there would be worn-out shoes and broken pots: 'marks of pastness' would be everywhere: at testimonial dinners there would be speakers in the middle of long speeches, whose hearers would be just as fatigued as if they had been listening for hours. And, in particular, historians would be at their work: in some five-minute old archive a five-minute old historian would be sifting five-minute old documents, and drawing inferences about events that never happened. There is, or rather, was, no past for their inferences to be about. Yet their behaviour is totally unaffected by this fact, for they *think* there was a past. But if their thinking in this way is wrong, and nothing they do is affected by this fact, what need have we for the concept of an actual past? What difference does it make if there was one or was not? We have so described the situation that there is no difference.

There is no difference, for instance, so far as carrying on with people whom we feel we have known for ages, but who, in fact, we never saw before: the man returns from the office to his wife from having left 'that morning', but she has no difficulty in recognizing him. As H. H. Price writes:

What matters . . . is not what my past actually was, or even whether I had one; it is only the memories I have here and now which matter, be they true or false. I recognize something here and now as being red. In actual fact, we are supposing, I have never seen anything red before. But what of it? I still have all my memories, erroneous as they are. Amongst them are memories of red things, and that is enough to make me recognize this one.[2]

Consider, in this context, thinking machines. These get stocked with 'memories', and, on the basis of these, the machine is able to carry out certain tasks. When the tasks are done, the machine's memory device is cleared, and new memories are fed into it. The machine never experienced whatever it is that these are memories of, but pragmatically this makes no difference at all. It uses its memories just the same way, whether they are true or false. On the present argument, we might think of the world as having been made five minutes ago, and stocked, as it were, with memories, or with things that function very much as memories do. There are in it, for instance, libraries. There are copies of Gibbon with footnotes, referring to other books, also in the libraries. So we can check up on Gibbon, remove discrepancies, offer reconstructions different from Gibbon's based upon other documents which are not cited, and so on. We proceed in all this just as we would if there had in fact been a Roman Empire which declined and fell, finally, at the time of Rienzi. But there *was* no such Empire. Nevertheless, the work goes on.

The distinction between memory and imagination is paralleled by the distinction between history and fiction. But in such a world as the one we are discussing (which could easily enough be just *our world*), these distinctions would, for the main part, be without basis. Unbeknownst to themselves, our historians would be writing fiction in a laborious way. Nevertheless, we would still distinguish between history and fiction, as between memory and imagination, just as we do in fact. A child might claim to remember having seen a bear yesterday, and his mother tells him he only imagines he saw a bear. Perhaps she persuades him. But if the world is only five minutes old, her memory goes back no further than his. What might make us say that she remembers and that he but imagines, is that her claim squares, as his does not, with the available 'evidence'. In her account things fit together which, in his account, do not. One might say, then, that it is this fitting together which gives us our criterion of truth: things which don't fit with what we accept, we then regard as false. But someone might now say: this *is* just how we operate. By fitting things together, accepting those propositions which cohere with what has antecedently been accepted, and rejecting those which do not. Notice, if we accept this, how natural *now* is it to say: statements purportedly about the past are really, so far as their cognitive significance is concerned, rules for predicting the outcome of historical research. We accept or reject historical sentences in accordance with whether they lead us to find further evidence. They enable us to organize what we find in the present world: a document takes us from the Colosseum, which we can experience now, to the Palazzo Farnese, which we can similarly experience now: and there we find the stones missing from the former. The statement 'The Farnese family took stones from the Colosseum to build their palace' serves to organize the two heaps of stones. Certainly there cannot be a question of comparing this statement with what it is ostensibly *about*. And it makes no difference whether there was or was not something for it to be about. Both possibilities are compatible with the conduct of historical research. We find the missing stones. But the world, perhaps, was made five minutes ago, with certain stones in the Palazzo Farnese which are congruent with certain holes in the Colosseum.

It is this that I find so deeply disturbing about the weird argument that the world might have been made, intact, but five minutes ago, everything being just as it is, just as it would have been had the world been as old as we believe it is. It is not simply that I should be disturbed if there were no way of proving it false. It is rather that it seems to make so very little difference whether it is false or not. But then the concept that it challenges seems to be far less important than one would naturally have thought it was. If the entire concept can be given up, this leaving everything else as it was, it seems hardly to be a concept which has any very significant role to play in our general conceptual scheme. And if the sceptical argument here has just the result of showing this, it has shown a great deal. For it is a great deal to have shown that a concept, heretofore considered of some importance, is of very scant importance indeed. I am not, of course, suggesting that there would be no *psychological* differences here. Something, one feels, might very well go out of life were people seriously to suppose there was no past. There would perhaps be little point in carrying on as historians now do, sifting evidence, etc., if there were nothing for the statements they arrived at to be about. Nor would there be much point in building up cases against defendants accused of crimes which could never have transpired, of which they could not in fact be guilty, even though everything 'fits' so that, if there *were* a past, we would say that they were guilty. There would, perhaps, be vast psychological differences. But here, a sceptic might urge, is one more instance of how much weight is put by human beings on what might, in the end, 'for all they know', prove to be but a fiction. Like, for example, their belief in a god.

The argument may make no difference in our lives, but there is something odd about it, and if we could identify the manner in which it is odd, we might be able to see what, if anything, is wrong about it. One way of making a start in this direction is to consider, for purposes of the contrast it affords, the symmetrical supposition that the world might be *annihilated* five minutes hence. The first thing to notice is that this supposition cannot in any obvious way be regarded as *sceptical*: that there will be no future does not sound of a piece with such propositions as that there was no past, or there is no external world, or that there are, perhaps, no other minds. Why the abrupt disappearance of the world seems feasible in a way in which the abrupt appearance of it does not, is perhaps not very easy to say, but the supposition, while pessimistic, does not seem sceptical, and it is one we have very nearly learned to live with. And one reason why it does not seem philosophically puzzling may be that it, unlike its symmetrical opposite number, does not clash with our notions of reference—a statement 'about' the future does not seem to refer in quite the same way that a statement about the past or present does—nor does it, in quite the same way, clash with the common use of ordinary temporal words. It is odd, for instance, to suppose that everyone in the world, a few very recent births aside, is of exactly the same age, namely five minutes, that most of the things there are have existed for just the same brief length of time. Yet there is no corresponding oddness in supposing that each person, however young or old, has exactly five minutes more to live (except a few who may die sooner): the hot lavae of Pompeii devoured young and old indiscriminately. It is, moreover, not nearly so odd to suppose that Levittown and Carcasson will perish together just five minutes from now, that each city will endure for just the same length of time, as it is to suppose that every city *has* endured the same length of time, and indeed just for five minutes. Again, we can easily suppose that no-one, except a very few persons fortunate enough to have children born to them within the next few moments, will have descendants, though it is hard to suppose that no-one, unless he has been born within the past few moments, has any ancestors. There is, not normally at least, anything which has the same sort of relationship to future events that memory has to past events, for example, precognition. But meanwhile it does not seem odd (largely because precognitive claims would themselves sound odd), to suppose that all the events precognized as taking place after the next five minutes will in fact not take place—though it is very odd to suppose that none of the events remembered as having taken place earlier than five minutes ago really did take place. And, finally, it is in no way odd to suppose false all those books which pretend to write the history of the next hundred years, because, to begin with, there are few if any such books and we should in the nature of the case expect them to be false. But it is odd to suppose that all the books pretending to exhibit the history of the past hundred years are false, for there are very many such books, and, in the nature of the case we would expect them to be true.

One could go on multiplying asymmetries and dissonances forever, but it does seem plain that the non-future possibility appears to involve none of the conceptual revisions enforced by the non-past possibility. I don't mean we would be unaffected if we were to take the former *seriously*. It would be a cruel blow to fond hopes, to plans, ambitions, and projects. It would terrify us, as the prospect of sudden death does. We are seldom, I think, as concerned over the fact that there was a time when we didn't exist as we are over the fact that there will be a time when we won't exist. I should be frightened were someone to tell me I had but five minutes to live, but should merely be puzzled were someone to tell me that I had lived for only five minutes. It offends me intellectually, but not practically. Practically, I might indeed say: what difference does it make, after all? The corresponding supposition about the future disturbs me practically, but not intellectually. I should have to be very stoic to say: what difference does it make, after all? It is difficult to accept, but easy to believe. Its opposite number is easy to accept, but, for reasons not yet clear, difficult to believe.

Now it is not enough just to register the fact that certain suppositions are odd, that they lead to the sorts of conceptual tensions which we have been able to reveal by showing that corresponding tensions do not arise on a symmetrical supposition. One wants some kind of explanation, and it seems to me easy enough, in at least a rough way, to account for the fact that we are able to accommodate the non-future supposition to our ordinary ways of thinking and speaking, and why it is with conceptual equanimity that we can tolerate the idea that the world and our way of viewing it should be just the same, though, just five minutes from now, the entire world will cease to be. It is in part because the future is not considered to have any effect on the present to begin with, and that the present is not causally dependent upon the future, inasmuch as effects do not precede their causes in time. By contrast, if we are to employ causal terms at all, the present is very much the effect of the past. Now these facts, if they are facts, are at the very least generally believed to hold, and they are certainly reflected in the language we employ for describing the world. Just to apply certain terms, and certain expressions, to present objects, *logically* involves making reference to certain past objects and events *causally* related to the object to which the term or expression is applied. Or better, let us effect a partition of the expressions and terms in our language into three classes, the members of each of which are normally applicable to present objects and events: (*a*) past-referring terms; (*b*) temporally-neutral terms; and (*c*) future-referring terms. For the present, I shall restrict my discussion just to (*a*) and (*b*).

By a past-referring term, I shall mean a term, whose correct application to a present object or event, *logically* involves a reference to some earlier object or event which may or may not be causally related to the object to which the term is applied. Again, I shall restrict discussion just to the causally related objects and events referred to by past-referring terms. A temporally neutral term, when applied to a present object, makes no reference to earlier or later objects or events. Let us now consider three distinct objects O-1, O-2, and O-3, under the temporally-neutral descriptions 'is a man', 'is a whitish shiny mark', and 'is a cylindrical metallic object' respectively. The criteria for applying these terms are specified with respect to certain manifest properties of the three objects, in the sense that one can tell, by simple inspection, whether or not these terms really apply to the object in question. Now consider, as applying to just the same objects, in just the same order, the three descriptions 'is a father', 'is a

71

scar', and 'is a cannon placed here by Francis the First after the Battle of Cérisoles in 1544'.

(1) The term 'father' is temporally ambiguous, in that one of its uses is temporally neutral: we appeal to essentially sociological criteria when we apply it to men. Yet this is not its primary use. Someone may, in the temporally neutral sense, be a father, and we might still want to know if he is *really* the father of the individual with regard to whom he carries out the socially appropriate paternal behaviour. As we know, a man may *not* be a father in the sociological sense and nevertheless be a father in the primary sense, as Talleyrand was the father *of* Delarcoix, though he never played the role of father *to* Delacroix. To be a father in the primary sense requires that, roughly nine months before the birth of a human being, the individual so-called impregnated the mother of that individual. Here, of course, the word 'mother' is not being used in the temporally neutral sense of 'mother to', but in the past-referring use 'mother of', that is, that the woman so called actually gave birth to the individual whose mother she is, in the way Jocasta was mother of Oedipus but never, or not always, mother to Oedipus. Correctly to call someone a father in the primary sense logically involves reference to an earlier event causally connected, in accordance with known principles, to the present. One cannot tell, by simple inspection, whether O-1 is a father in the primary sense. One can infer, of course, that O-1 is a father, on the basis of other properties of O-1 which can be seen to hold on the basis of simple inspection.

(2) The predicate 'is a scar' is temporally unambiguous. If O-2 was not caused by a wound, it just simply is not a scar. It is only scar-like. Correctly to describe something as a scar, then, involves, logically, a reference of some earlier event which stands, to the object so described, in some obvious causal relation. If whitish shiny marks were to appear spontaneously, like stigmata, upon ones body, one would describe them as scar-like, but not as scars. 'Scar-like' is temporally neutral, *unless* we understand it to make a *negative* reference to the past, namely that it was *not* caused by a wound. In this sense, 'scar-like' differs from 'father to' in that the latter makes no reference, positive or negative, to the past. He who is father to *x* may or may not be father of *x*.

(3) The third description makes an *obvious* reference to a past event, and

had there been no such past event, the description itself would be false, or legendary, like 'the rock placed here by the Titans after their victory over Uranus'. The only interesting difference between this case and the other two is that there are no obvious causal laws connecting cannons in St-Paul de Vence with actions of sixteenth-century French monarchs. True, one might say that the cannon had to be placed here by *someone*, but in fact it is not clear even that the cannon was *placed* here: it might just have been left here. And this of course determines whether it is to be called a *monument to* the victory or merely a *memento of* the victory.

It seems to me that temporally-neutral predicates are logically independent of past-referring predicates, and indeed I have tried to define them that way. But I don't think I am merely legislating here: it seems plain that something may be a man and not be a father, be a white shiny mark and *not* be a scar, be a cylindrical metallic object and *not* be a cannon, much less a cannon deposited by Francis the First who once employed it. By contrast, past-referring predicates are *not* independent of temporally neutral predicates. Nothing can be a father which is not a man, etc. The compositional relations amongst the two classes of predicates is complex, and the philosophical problems are comparable to those which arise in connection with relations between other classes of terms, for example 'is an arm movement' in contrast with 'is a gesture of farewell', or 'is beautiful' in contrast with 'is red'. Just now all I am concerned to stress is that there is an interesting analogy between temporally neutral terms and past-referring terms, on the one hand, and tenseless and tensed sentences on the other. For a tensed sentence seemed to presuppose, for its truth, a tenseless true sentence. And comparably, for a past-referring predicate to be true of a present object, some related temporally neutral term must be true of it first. We can falsify 'is a father' by demonstrating that 'is a man' fails to apply. But we cannot be certain that 'is a father' applies simply because 'is a man' does.

Our language is saturated with past-referring predicates, and one might plausibly suppose that Lewis's notion of marks of pastness was based upon a not uncommon philosophical tendency to mistake a structural feature of our language for some structural feature of the world, and, in his case, to look to some mysteriously absent properties of things as what we must be referring to when we use this part of our language. But we are not referring to present properties of things when we use past-referring predicates, though in some sense our application of these terms to present objects *does* depend upon the object having certain properties which can be seen to hold on the basis of simple inspection. Rather, we are referring to certain past objects and events. The house in which George Washington slept looks like a perfectly ordinary house, and there is no special property of it we can look for in order to determine that it was slept in by the First President. There are, if you like, no such properties, or none at least that we can notice on the basis of inspection. The criteria for applying them are rather more complicated, and the decision as to whether they are *true of* the objects they are applied to are more complicated still. Whatever the case, since past-referring predicates, when true of present objects, give us information about events and objects which are *not* present, it is plain enough that we cannot fully translate into a temporally neutral idiom sentences which employ these terms. For a full translation of a sentence S into a sentence T must, in addition to preserving the truth value of S, convey the same information that S does. If *untranslatability* of one set of terms into another set of terms is our criterion for distinct *levels* of language, then the two classes of terms here are of different levels though they apply to the same things, viz. O-1, O-2, and O-3.

Now it may be said that the non-past possibility we have been concerned with does indeed leave unaffected the level of language which uses only temporally neutral expressions. All these predicates are true of objects whether there was a past or not. But this can hardly be said of predicates on the other level. It is not simply, on the non-past possibility, that all statements purportedly about the *past* are false. It is also the case that a great many sentences about the *present* would be false as well, all those sentences, namely, that ascribe past-referring predicates to present objects, objects which would still have all those properties, the presence of which we can discern on the basis of simple inspection. Our two sets of terms are made up of extensionally equivalent pairs, in that each term in one pair designates exactly the same object that the other term in that pair does. But one member of each pair presupposes, for its application, some fact about the past. So everything would be just as it is.

74

Only there would be no fathers, no scars, and no cannons left or placed by Francis the First. But nothing would have disappeared: there would still be all the objects currently designated by those terms, that is, men, shiny and white marks, cylindrical and metallic objects. And not merely would all such sentences be false because the predicates they employ are false of the objects they are applied to but also, all the causal laws presupposed in the use and application of most of our past-referring predicates would be false or, if not false, then vacuous.

Turning now to future-referring predicates, the main thing that must impress us is how hard it is to find any natural examples, if, by such a predicate, we mean one which refers to some *future* event or object, as a condition for applying to some present object or event. Consider the predicate 'is a father-to-be' as applied to the consort of a currently pregnant woman. To be sure, we customarily expect that there will be a child, and that the man will be a father, 'if all goes well'. But in fact we apply the predicate 'father-to-be' on the basis of either temporally-neutral or past-referring predicates, which we suppose hold true of the individual so designated. Thus x is father-to-be in the case where x has impregnated y and y has not yet delivered the child. And nothing more is required. If x should die before the child is delivered, or if y should, or if the birth is aborted, still, x was a father-to-be. It is not required that he should become a father afterwards. His title as father-to-be does not logically depend upon what the future brings. Moreover, our expectation that x will be a father if he is in fact a father-to-be is based upon causal laws which *have* held, and such future-referring predicates as we might ordinarily use would then be parasitic upon our ability to use past-referring predicates, since the future, in regard to causal laws, 'must resemble if not reflect' the past. The main point, however, is that what seem, on the face of it, to be future-referring predicates are for the most part readily translatable into past-referring or temporally neutral language, and their application to present individuals does not *require* any later occurrence. So if the world were to end five minutes hence, none of the sentences which use such predicates in descriptions of present objects would be false. If this is so, then, the truth of no sentence about the present presupposes the truth of any sentence about the future, and this, if so, would explain why we find no difficulty in accommodating to our conceptual scheme the idea that there might, very soon, be no future at all.

There are, of course, some descriptions of past events which, had they been given at the time the events themselves took place, or even before then, would have *had* to make use of future-referring predicates. Now we may refer to Piero da Vinci as the father of the man who painted *La Gioconda*. To have called him that when he was father-to-be of Leonardo would logically require that his child come to paint *La Gioconda*. Here *would* be a description of a present object whose truth would depend upon what the future brings, and one, moreover, which could not in any obvious way be translatable into either past-referring or temporally neutral expressions. For the required painting did not yet exist, and the description would give us, if true, genuine information about the future, and so it could not be translated into expressions which did not give that piece of information. But when one thinks how odd it would sound to hear such a statement being used, in comparison with the non-oddity of the non-future possibility, one gets some idea, I hope, of what I earlier meant when I spoke of substantive philosophers of history talking about the future in ways ordinarily used only to talk about the past. But the explanation of this oddity must be reserved for a later discussion.

None of these considerations, of course, affects the sceptical argument that, for all we know or can know, the world might have begun five minutes ago, and that at the very least such a claim would be logically possible. It does not affect it because the use of past-referring terms presupposes certain theses regarding causality, and the sceptical argument is precisely an attack on certain notions of causality, an attack, in a way, which goes back at least to Hume, whose point was that causes do not logically entail their effects, that from a description of the manifest properties of one thing, we could not logically *deduce* what effects it would have, nor, from an exhaustive description of another thing, could we deduce what its causes must have been. Our causal concept is built up out of certain associations with respect to what has *in fact* happened, but there is nothing compelling, logically at least, about such associations, and the presence of a given thing is logically compatible with having had different causes than it in fact had or, for that matter, with its having had *no* causes at all. Hume writes:

When we exclude all causes we really do exclude them, and neither suppose nothing nor the object itself to be the cause of its existence; and consequently can draw no argument from the absurdity of these suppositions to prove the absurdity of that exclusion. If everything must have a cause, it follows, that, upon the exclusion of other causes, we must accept of the object itself or of nothing as causes. But it is the very point in question, whether everything must have a cause or not; and therefore, according to all just reasoning, it ought never to be taken for granted.[1]

In a way, my discussion has but extended Hume's idea that we cannot, from an exhaustive description of the manifest properties of things, deduce their causes. My extension consisted in showing the irreducibility of past-referring predicates to temporally neutral ones. So far I have only tried to show that this, taken together with the fact that all natural predicates which seem to refer to the future are in fact eliminable in favour of temporally neutral or past-referring terms, accounts for the ease with which we can accept the possibility of a non-future, and the corresponding unease that the suggestion of a non-past induces. But this is no proof that there is as yet anything wrong with the non-past hypothesis, if we take it as such. For the use of past-referring predicates presupposes that things in the present world have had causes in the past, and it is precisely this which is in issue. We can hardly defeat an argument by merely presupposing what it attacks. And basically what it attacks is the idea that there *is* some logical connection between events or things, and it is this that makes the argument a logically possible one:

There is no logical impossibility in the hypothesis that the world sprang into being five minutes ago, exactly as it then was, with a population that 'remembered' a wholly unreal past. There is no logically necessary connection between events at different times; therefore nothing that is happening now or will happen in the future can disprove the hypothesis that the world began five minutes ago. Hence the occurrences which are *called* knowledge of the past are logically independent of the past; they are wholly analysable into present contents which might, theoretically, be just what they are even if no past had existed.[2]

The italicized 'called' might be explained in this way: The context

'. . . knows *a* . . .'—where 'a' denotes anything you choose—entails that *a* exists.[1] If, then, someone could correctly be said to know the past, this would entail the reality of the past, and it *would* be incompatible to both assert that someone knows the past and to say that what he knows about never existed. A comparable point might be made about scars. Given present rules of usage, to say that someone bears a scar entails that he once suffered a wound. Similarly, in the case of knowledge, we must speak, rather, of what is *called* knowledge (but isn't or at least may not be), just as, in the case of scars, we might rather speak of what are *called* scars, but are not (or may not be). I think enough has been said to suggest that we cannot analyse 'is a scar' into temporally neutral language, into 'present contents', so what we had better say is that if the hypothesis Russell advances is correct, every description of shiny white marks as *scars* would be *false*.[2] But comparably, every description of cognitions as *knowledge of the past* would similarly, and for similar reasons, be false. The world, in short, would be just as it is. But our language for describing it would be different.

These remarks have had, I think, the positive result of showing that our notion of the past is connected with our notion of causality, and that our notion of causality is connected with our language. As a bit of psychological speculation, I should like to suppose that children begin with a temporally neutral language, and then, at the same time, so to speak, acquire together the use of past-referring terminology, a concept of causality, and a concept of the past, all three achievements being interdependent. It would only be natural, then, that any attack on our concept of the past would at once involve an attack on the concept of causality and upon our use of past-referring terms.

'Like all sceptical hypotheses,' Russell wrote, 'it is logically tenable, but uninteresting.'[3] We have seen, I think, that on the contrary it has a considerable interest. Whether or not it is logically tenable remains to be seen.[4] It seems to me possible to offer a certain analysis of the role which sentences purportedly about the past play which the non-past hypothesis merely serves to dramatize. The analysis, roughly, is this. Sentences purportedly about past objects and events are not, as we have seen in earlier discussions, properly to be understood as *about* the evidence offered on their behalf, nor are they capable of being fully analysed into sets of observation sentences; in fact, the truth of any set of observation sentences fails even to be a necessary condition for the truth or falsity of sentences purportedly about the past. Nevertheless, such sentences might function, in historical inquiry, in a role analogous to the one played by sentences employing so-called theoretical terms in science, and stand to observation sentences in just the same sort of relationship that those sentences do. And one might now say this: their role is chiefly one of serving to organize present experience. If this analysis were correct, the question whether or not they were independently *about* anything would fail to arise, and the non-past hypothesis would then be irrelevant, and has served only to draw our attention to a mistaken notion we have had regarding the function of these sentences in the economy of human cognition. A term like 'Julius Caesar' enjoys, in historical work, somewhat the same role that 'electron' and 'Oedipus Complex' enjoy in physical and psychoanalytical theories respectively. Sentences employing these latter terms do not stand or fall on the question of whether or not they denote actual entities, albeit unobservable ones. For they would play the same role in the organization of experience whether they did this or not. It is well known that there is a problem, which has been solved only in a trivial and unacceptable way, of eliminating theoretical terms in favour of mere observational vocabulary.[1] Yet in using sentences which embody them, we are not thereby committed to allow unobservable entities. There may or may not be such entities, but it makes no difference whether there are or are not. Their essential role is unaffected by the issue of denotation. Such sentences, as instruments, need no more admit of truth-values than do other scientific instruments, for instance test-tubes. And these sentences, like test-tubes, are indifferently available to scientists who may otherwise differ on questions of ontology—a differing which is, if you like, a luxury of intellect which has no bearing on their use of theoretical vocabulary in the organization of experience.

I shall call this analysis the Instrumentalist view of sentences about the past. Instrumentalism, of course, is but one of a number of possible positions which have been taken with regard to theoretical terms. A full-scale discussion of the problems involved is quite beyond the scope of this work, and belongs properly to the philosophy of science. No

Instrumentalist I know of has ever extended his favoured analysis of theories to *historical* sentences,[1] but it seems a natural move in the present context, if only for purposes of neutralizing the force of the sceptical argument and, incidentally, of pointing out a plain analogy between theoretical science and history—an analogy often disregarded in discussions which *contrast* historical and theoretical science. It is an analogy which will later stand us in good stead, whatever stand we take ultimately on Historical Instrumentalism as a *general* analysis of historical sentences.

I think, however, that as a partial, functional analysis of these sentences, Historical Instrumentalism is almost certainly correct. Historical sentences *do* play a comparable role to theoretical sentences with regard to organizing the present. We find, for example, a pair of plays which exhibit, let us suppose, certain striking stylistic similarities. By postulating a single author for them both, we organize these works into a single corpus. Similarly, we find marked stylistic discrepancies in a pair of works considered as forming a single corpus, and postulate *different* authors, reorganizing, in this way, extant literary works. We then proceed to look for further parts of the present world to support these different organizations, and so relate, once more, parts of the present world to other parts. We can regard these as theories, I think, with no great difficulty: we may speak of the Single Author Theory and the Double Author Theory, and allow that such theories serve, *inter alia*, to organize the observable world.

Notice, however, that this notion of a theory does not *rule out* the possibility of such a theory being true as well as useful. Driving, I notice that the indicators show the car to be overheating, and the battery to be discharging: two red data. I proffer the theory that the fan belt is broken, since this would account for the fact that the battery is not charging and the car overheating. This serves, doubtless, to organize my readings off the dash-board, but it is a theory which collapses into a *fact* when, peering under the hood I detect a broken fan belt.[2] Is it simply the lack of access to the past that prevents historical theories from similarly collapsing into facts? For one cannot but feel that one difference between historical theories and the sorts of scientific theories we have been connecting them with is that, while the latter have reference to what, if they were entities at all, would be singularly different from the entities to be encountered in gross observational experience, the entities postulated by historical theories are exactly of the sort encountered in everyday life. That is to say, no one has in fact observed such things as atoms, electrons, Psi-functions, genes, and ideas surcharged with libidinal energy, but the everyday world contains, amongst other things, authors. So historical theories make use of terms which have a plain application to things which are presently capable of experience. It is not a difference in the kind of entity postulated from everyday entities which then makes the difference, but merely the epistemic inaccessability to historical entities which has encouraged the move to Historical Instrumentalism.

Now it might be countered that I have shifted ground, and moved from considerations having essentially to do with the concept of causality, to considerations which have essentially to do with knowledge. But the force of the non-past hypothesis derives its force from the fact that we have, apparently, no epistemological access to the past, and hence no independent way of checking up on it. If we could have such access, we would have a means of collapsing theories into facts, and at the same time ways of empirically refuting the non-past hypothesis. It would then be an empirical hypothesis and nothing more, and subject to empirical falsification. So epistemic considerations are surely not irrelevant. Once we introduce them, however, we can map a strategy for handling the non-past hypothesis.

To begin with, we retreated to instrumentalism[1] as a way of neutralizing problems of reference which arise in connection with sets of statements, in this case statements about the past, whose referenda were deemed inaccessible even if they once existed. Instrumentalism proposes to circumvent all questions of reference by showing that it does not matter whether they refer or not: everything would remain the same, but we would only have converted certain sentences from fact-stating to organizational instruments, and in the latter capacity truth or falsity are rendered logically inappropriate: there are only 'better or worse' such instruments, as Dewey would have said, the latter values being functions of the relative organizational achievements of pairs of sentences. Yet it is possible to manufacture an indefinite number of *ad hoc* scepticisms, each of which could be circumvented by a similar retreat to Instrumentalism. Suppose, for instance, that someone were to offer the hypothesis that the world ends exactly five feet beyond ones furthest reach:[2] that just five feet from wherever one stands, there is nothing, so that statements purportedly about the Empire State Building, made by someone standing in Central Park, would be false for lack of the object referred to. We speak of things as thirty miles distant from *here*, but we have no access to these, and someone might then suggest that we adopt a spatial instrumentalism, avoiding thus problems of reference by relegating 'Empire State Building' to the status of a theoretical term, so that sentences embodying it serve to organize spatially accessible (observable) phenomena.

The fatal difficulty in such scepticisms is their sheer arbitrariness. Why is the line drawn where it is drawn, and not somewhere else?[1] Why draw the line at five feet and not at six or at four feet? Or seven or three? Why five minutes ago and not four, or three or six or ten? Or, for that matter, if one wants to say that objects five feet beyond our furthest reach are inaccessible, because we cannot touch them, and that, for all we know, they are not there, why not say that we have no way of knowing, at the present moment, that there is anything at all except what we are now touching? Or that there is anything at all except what we are seeing *now*? One might suggest that though we are not touching them, we *can* touch them, and know they are there. But then why cannot we move and touch the things now beyond our reach by five feet? You cannot say that they are not there, that five feet away there is nothing, for this quite begs the question, namely, how can we know whether there is anything there or not? The thing is, we are where we are, and not five feet from that place. But, for that matter, we are touching what we are in fact touching, and not something else. So that to suppose continuously tangible entities between touch-events is, if you like, to introduce theoretical entities for purposes of organizing experience, and continuous physical objects serve to show that here, too, we have retreated to a new instrumentalism. It should be plain, then, that these differential scepticisms rather quickly collapse into scepticism per se, and talk of objects in general is cast in the instrumentalist mode.[2]

In my earlier remarks upon the non-past argument, I pointed out that it is not perfectly general, that it does not rule out all statements about the past, but only those which purport to be about something having existed or taken place more than five minutes ago. But I have now emphasized how arbitrary is the specification of five minutes. There are innumerably many other points at which the line could have been drawn, and, if all the evidence is compatible with a world five minutes old, it is compatible as well with one six or seven or *however* many minutes old. All the evidence is, if you wish, compatible with infinitely many hypotheses, each of which is incompatible with the others. But to justify drawing the line at one point rather than another can surely only be done by some appeal to evidence, and if appeals to evidence are ruled out, there can *be* no possible justification for entertaining one hypothesis rather than another, for instance that the world sprang into being five minutes or five years or five centuries ago. Every differentiating piece of evidence can be neutralized by an adherent of a shorter time span that the one it purports to establish.

Let us now opt for the very hypothesis that the world sprang into being five minutes ago. Notice that this effects a partition amongst the class of statements about the past. If the world began five minutes ago, some statements about the past are true or false, namely, those statements about what have transpired in the past (the only) five minutes.[1] The rest, lacking referenda, are either false, or else their truth or falsity cannot arise, or else questions of truth or falsity are irrelevant since these statements have the status of theoretical sentences. Let us continue with historical in- strumentalism, and say that some statements are to be analysed instru- mentally, and some not, the latter being about what has happened very recently. Notice, also, that there will be a corresponding partition within the class of past-referring predicates. We can admit those past-referring predicates which refer to past events and objects connected with presently existing objects, so long as they happened or existed within the past five minutes. Thus there would genuinely be three-minute eggs, and not merely eggs *called* three-minute eggs. There would be *some* fathers, *some* genuine memories, and so on. But now, as we shift the arbitrary begin- ning point back and forth in time, the populations of these various classes will vary. If we move it far enough back, everyone ordinarily called a father will be a father, and there will be genuine memories, genuine scars, and, indeed, cannons genuinely placed by Francis the First. And more and more statements will really be about the past and not merely be useful instruments for organizing the present. As we move it closer and closer to the present moment, however, there are fewer and fewer genuinely past-referring predicates, and fewer and fewer statements genuinely about the past. We reach a point where the only genuinely applicable predicates are temporally neutral ones, and the *only* role left for statements purportedly about the past to play consists in organizing the data of the present. There is room for nothing *but* Historical Instrumentalism. But is there any good reason why we should not reach that point, why we should not bring the origin of the world closer and closer until, finally, there just is no past at all, not even a five-minute one? Is there no evidential friction from keeping these temporal scepticisms, of which, as I have said, there is an infinite number, from sliding into *instantaneous* scepticism? The answer is that there is not. For no better evidence can be given for one rather than another, even though, as I have suggested, each one of them allows *some* genuine statements about the past. Since they cannot justify this allowance, however, their granting it hardly matters.

I do not believe that instantaneous scepticism is ultimately tenable at all. There is a clear, and analytically true sense in which one might say that only the present exists. It follows from this that the past does not exist, but this amounts to little more than the triviality that the past is not the present, and hardly entails that the past *did* not exist. Moreover, it is not clear that when we speak of the present, we are speaking of an *instant*. When we point to something and say that it now exists, we are not saying, so to speak, that its existence is confined within the present instant, for an instant has no confines within which something may exist. An instant is no more a unit of duration than a point is a unit of extension. The spatial analogue to instantaneous scepticism is, I suppose, punctiform scepticism. We can, however, hardly speak of things having punctiform existence. This would require that a circle have its centre and circumference coincident with one another, and this simply disqualifies it from being a circle. Punctiform scepticism entails that nothing exists; this *is* pure scepticism. But so, too, is instantaneous scepticism just plain scepticism. To be a thing is to have extension and duration, and to deny either of these is to deny the existence of things.

We do, of course, recognize these days the occurrence of achievements,[1] in so far as we accept Professor Ryle's important distinction between achievement verbs like 'winning' and other verbs like 'running'. Running a race takes time, but winning one does not: one wins *at* a time, but not through an interval. But the point of the distinction is lost if every verb is an achievement verb, and surely one must run a race in order ever to be said to have won one. Moreover, it is runners who win races, and runners are entities, and so have some duration: there are instantaneous winners (every winner is one), but no instantaneous racers. An instant marks a temporal position, is a device for calibrating time, but is not, I think, part of time, nor does it have times as parts of itself. Instants do not belong in the list 'year, month, week, day, hour, minute, second'—any more than points belong in the list 'mile, rod, yard, foot, inch'. This can be seen from the fact that nothing can endure two *instants*, though it can endure two hours, minutes, or seconds; nor can anything extend over two points. There are no points if there are no lengths, and no instants if there are no durations. So in a sense, to speak of instants presupposes durations, and one cannot accordingly adopt an instantaneous scepticism and hope, by doing so, to raise doubts about durations.

These considerations, if sound, entail that one cannot coherently maintain instantaneous scepticism. My arguments have not, in any way that I can think of, presupposed anything concerning causality. One can quarrel, if one wishes to, over the precise duration the world is said to have, but not over the question whether the world has any duration at all: just to be a world at all requires that it have *some* duration, and the only remaining question is how much. The man who wishes to claim that it has five minutes' duration only, as we have seen, is committed to saying that some statements about the past are true, those which have referents falling within the temporal restrictions he has imposed. But now we may ask him how *he* knows. This is no longer a problem he can escape, for if he says only that his chosen duration is arbitrary, we can stick him with instantaneity. The point is, he cannot opt for his choice without allowing something to count as evidence, and if he allows any why not all? The issue over the duration of the world is an empirical issue, decidable in principle, and if it is not that, it comes to instantaneous scepticism, and *that*, I am saying, cannot be held. It cannot be held because it is self-contradictory to hold it. One cannot speak of an instantaneous world. It is comparable to asserting that there was a past but we cannot know there was one. The only position left to occupy then is plain scepticism, and it is not our task to discuss this, for plain scepticism raises no *special* problems for the philosophy of history.

This is as far as we can go, I think. Now to offer a hypothesis concerning the precise amount of duration to be assigned to the world is to be prepared to accept something as evidence for this hypothesis and against competing ones. But the fact that we must accept something as evidence for this proposition about the past brings us to the threshold of our third argument against the possibility of our succeeding in making true statements about the past. For it is just here that the relativistic factors supervene: statements about the past must be relative to bodies of evidence. Before crossing that threshold, however, I want to make one point more about the present argument.

As we revise our estimate of the precise amount of duration to assign to the world, we restore increasing amounts of our temporal vocabulary, and increasing numbers of what are accepted as causal laws. A five-minute world, as things stand, is too short a time for there to be any genuine scars. Suppose, now, it takes a month for an average wound to become a scar. Then to suppose the world a month old gives us some scars but no genuine fossils. To say the world is a million years old restores to genuine usage the past-referring predicate 'is a fossil'. The further back we go, the less strain is put on either our causal scheme or our temporal vocabulary, and if someone were to say: Suppose the world sprang into being a hundred million years ago, it is hard to see how we should find this very sceptical or even very interesting philosophically. It allows us all of history and a good deal of prehistory, and if he were to add 'intact with everything in it' we would hardly find this disconcerting unless we knew what the world was like at that time, and our so knowing would then create tensions with known causal laws and a temporal vocabulary in use. If the intact world contained, say, gravid dinosaurs, we should have to revise some causal notions if we were to accept this estimate. But the simpler the contents of the world, and the fewer temporal predicates required to describe it, the less jarring would the notion of intactness be. Even the story of creation does not require that the world sprang into being, but only that it was made, and then that it took six days to stock it properly. There is nothing logically absurd in the idea that the world was created, or that it was created however long ago one pleases. I have only tried to show that any such hypothesis, if arbitrary, slides quickly into absurdity. But not every such hypothesis is arbitrary, and only the possibility of empirical support prevents its logical collapse. With this we may finally turn to the fresh set of difficulties which the admission of evidence apparently imposes upon us.

VI

EVIDENCE AND HISTORICAL RELATIVISM

Historical Relativism, as a form of scepticism regarding our ability to make true statements about the past, stands in marked contrast to the two forms of scepticism we have so far considered. To begin with, it has been taken seriously, and has been actively endorsed, by a great many practising historians, including some very distinguished ones. And, in setting the position up, one apparently requires, in contrast with the first sort of scepticism, that some statements really are about the past, and, in contrast with the second sort of scepticism, that there really is a past for such statements to be about. That there really is, or was, a past, is insisted upon by Charles Beard, who speaks of 'history as past actuality', meaning, by this, 'all that has been said, done, felt, and thought, by human beings on this planet since humanity began its long career'.[1] We do not, of course, have direct access to history-as-actuality, but only indirect access through using 'history as record', that is to say, documents, monuments, symbols, memories, or bits and pieces of the *present* world which stand in certain relations with history-as-actuality. Finally, Beard speaks of history-as-thought. This is *about* history-as-actuality, but 'instructed and delimited by history as record'.[2] So far there is plainly nothing much out of the ordinary in this analysis:[3] indeed it *is* simply the ordinary way of looking upon the activity of the historian. Beard really departs from the ordinary account only by holding that certain causal factors, operating upon the individuals who seek to make statements about history-as-actuality, somehow deflect them from making true and objective statements. I shall, however, begin by concerning myself only with the ordinary part of Beard's discussion, and deal with the relativity factors afterwards, hoping to show that they are not so damaging as Beard and some of his supporters and critics have been led to suppose.

It is refreshing, after all this time, to deal with a view so soothing to our ordinary notions about history, a view which holds that we are after all logically enabled to make statements about the past, instead of statements which, though aimed at the past, are somehow always turned aside from their target and hit, instead, the present or the future. Beard's contemporaries, the Pragmatists, insisted that statements about history-as-actuality are to be analysed, in the end, as statements about history-as-record, as though the documents, monuments, and so on, stood as a statement-proof curtain between ourselves and a past we could not so much as mention. It is not the documents, however, which stand *between* us and the past, on the contrary, they are just what enable us to find out about a past we could not begin to chart without 'record and knowledge authenticated by criticism and ordered with the help of the scientific method'.[1] They provide the means but not the object of historical query. Ronald Butler puts this succinctly:

When we claim to know a past event, we are doing something different from merely evaluating the evidence. On such occasions, we do not think of the evidence as an impenetrable curtain: we claim to be looking through the fabric and beyond. ... We have yet to analyse 'looking through the fabric and beyond'.[2]

This is certainly a difficult notion to analyse, partly at least because—if I may protract the metaphor which does not quite permit one to say what one wishes—it is only by 'looking through the curtain and beyond' that we can see the fabric. Less metaphorically, just to apprehend something as *evidence* is already to have gone beyond the stage of merely making statements about *it:* to count something as evidence is already to be making a statement about something else, namely, that for which it is taken as evidence. And taking E as evidence for O is to see E differently from the way we would if we had no notion at all about O. Thus, just to see something as *evidence* is already to be 'looking through the fabric and beyond'.

With regard to statements about the past, to see something E as evidence for one of them is to be seeing E in a certain temporal perspective. And indeed, it is only with reference to the past that we can license certain descriptions we give of what we see. Of a traveller returned from Venice, I may ask whether he saw the *Rondanini Pietà* of Michelangelo. This is an instructive example, for, not very long ago, those who saw the same object which is now so-called would not have said they saw the *Rondanini Pietà*: until quite recently, this object was taken to be a detached piece of the foundation work in the Palazzo Rondanini in Rome, and those who saw it, saw it as *that*. Now, restricting ourselves to temporally neutral vocabulary, there are predicates true of this object that are compatible with both descriptions, for example, it is marble, is so many centimeters along its longest axis, weighs so many kilograms, and so on. To speak of it as a foundation stone is already to give it some temporal dimension, that is, to relate it to antecedent acts of masonry. Similarly, to speak of it as a statue is to relate it to antecedent acts of stone carving. To describe it as a statue by Michelangelo, or better, as the last of his four known pietàs, is to make use of a rather precise temporal predicate. It is because people did not see the pietà as something this predicate was true of that it ended up in the *palazzo* basement. The important thing, however, is that to designate it thus is to connect it with the past; to see something *more* than a piece of marble, so many centimeters long along its longest axis. To be able to see it as Michelangelo's work is to have looked through the curtain and beyond. Employing our temporal vocabulary, irreducible, as I have argued, to a temporally neutral idiom, is to be past the curtain. Or, if you wish, the curtain falls between past-referring language and temporally neutral language. But in temporally neutral language we cannot so much as speak of what we see as historical *evidence*.

Asked for a closer analogy, I would recommend that one think of objects in the present world as comparable to words, and the historical use and understanding of them as comparable to reading. It does not ordinarily occur to us to think of the words in which something is written as an inky curtain *between* us and the meaning they have: in reading, if you like, we see through the words and 'beyond', and indeed hardly ever notice the words in their status as physical objects, that is, as hooks and circles of dried ink. There are, I suppose, three main classes of individuals who see *only* marks when presented with a sheet or words: the illiterates, people literate in one notation but illiterate in another, and those who suffer from a certain sort of brain damage. The Sicilian peasant who does not see a certain stone pile as a Norman tower is historically illiterate: he does not know what the stones *say*. A classical scholar examining an Etruscan inscription, knows already that these marks are meaningful, but does not know what they mean: he has not learned to read these marks and, in a sense, is therefore abnormally concerned with the marks *as* marks. It is not easy to find an example of someone here who fits the case of the man with brain damage. The closest I can come to it is the philosopher who insists that *all* we see are present things, that we do not see the past; a person, in short, for whom reading is somehow an unintelligible activity. We do not see the past now; we see only what is before us. But reading is an interpretive activity, and just to see marks as words is already to see them as demanding interpretation. We cannot, surely, translate what a book is about into an indefinitely long set of statements about the letters it is printed in, unless we stretch our concept of 'letter' and so build into it the precise concepts we would have sought, through translation, to eliminate. It is in just this sense that I have been insisting that the language of time cannot be paraphrased in temporally neutral vocabulary.

Returning now to Beard's distinction, we might say that it is only with reference to 'history as actuality' that anything is to be constituted as 'history as record'. Accordingly, it is a shade disingenuous to wonder how it is possible to move from 'history as record' to 'history as actuality'. For just to be seeing something as 'history as record' is already to have made that move. Otherwise we would just be seeing *things*. Too frequently, epistemological discussions in history begin by assuming something that is false: that we are all of us temporally illiterate. It then becomes a pressing question how we can go from present to past. The answer is that we cannot. We could not for the plain logical reason, that in any inference from present data to past fact, we would certainly require some general rule, some principle, some proposition amongst the premisses of our inference of *this* form: 'If E then F' where E refers to a present datum and F to some past event. And it would be a baffling problem to account for such propositions if we had only what is present to go by. We could, indeed, not understand such propositions. We do understand them readily enough, however, and those who could make sense of nothing but temporally neutral language can hardly account for this fact. One may insist, in a kantian way, upon the categorical character of time. But whether this is correct or not, it is a fact, I think, that we automatically acquire our concept of a past as we acquire our language, which is rich in past-referring predicates. Without this, I cannot see how history should so much as begin. It was a great insight on Vico's part to regard the beginning of language and the beginning of history as of a piece. No one who is master of his language can, I say, live wholly in the present. Or he can do so only by a special wrench of the will, and is, because he must be *rejecting* a richer temporal existence, living in the present in only a derivative sense. Roughly in the way that Marie Antoinette was living an Arcadian life at the Petit Trianon at Versailles: a matter of *mauvaise foi*.

These considerations might be brought out in another way. It has been shown, I think conclusively,[1] that one cannot give definitions of our various temporal distinctions without making at least tacit reference, in the definientia, of the temporal notions already contained in the definienda. Someone, for instance, might try to define the past as that which it is logically possible to remember. But the use of the word 'remember' in the defining expression already makes some tacit reference to the notion pastness: it is part of the *meaning* of 'remembers' that what is remembered is past: 'remembers', as I would say, is a *past-referring term*. So such a definition would be contaminated by circularity. One might, comparably, undertake to define the *present* as that which one *is* in fact experiencing.[2] It follows that one cannot be experiencing the past or the future. But 'is experiencing' involves a relation between an individual and an experiendum, in particular a confrontation of the latter by the former, and one can only confront, in the relevant sense, what one is contemporary with. So the notion of presentness is, once more, already contained in the defining expression. True, one might go on to say that it is logically possible to experience both past and present, but this means, once one has analysed it, that one could have experienced something when it was present, or that one will be able to experience something when it will be present. The concept of presentness is built into the concept of experience, so that it is analytically necessary that one can only experience the present. Given this logical fact, together with an identification of knowledge with experience, it becomes difficult to see how one can know the past: the past is one of the things which is not experienced. But if we go on to ask how, if all we experience is the present, we can possibly know the past, we are assuming a temporally neutral sort of experience. And this, I have been arguing, is unrealistic. If we heed the manner in which we do in fact experience the world, the really pressing question would be this: how, if we did not know the past, could we possibly experience the present world as we do? For in fact, and our language shows this, we are always experiencing the present world in a logical and causal context which is connected with past objects and events, and hence with reference to objects and events which we cannot be experiencing at the time when we are experiencing the present. And there is no way of widening, so to speak, our concept of the present so as to bring these into the present. When Proust experiences the tea-steeped *madeleine*, he is not also *experiencing* Tante Leonie and Combray: he is remembering them with a remarkable clarity, and the whole force of the passage in which he describes all this depends upon our recognition that these things are not present and will never be, that they are irremediably past, and can only be artistically recaptured. But we do not recapture what we experience at the *time* we experience it: 're-captures', in the specific Proustian sense, is logically a past-referring term. His experience with the madeleine is a dramatized example of what happens typically and commonly on the part of those who have acquired any past at all. His sense of *puzzlement*, just after tasting the cake and tea, about what these *mean*, is special, of course: it is a case of not being able momentarily to read something that one knows how to read. But it is for all that, experiencing the present as significant of the past without being able quite to say what it is that it is significant *of*. In fact he has experienced the present in the light of the past without, for a difficult moment, being able to establish a connection, until finally it becomes clear that 'the taste was that of the little crumb of madeleine which on Sunday Mornings at Combray . . . when I used to say good day to her in her bedroom, my Aunt Leonie used to give me . . .'. In a less personal manner, of course, and without any comparable reference to memory, all of us might sense the same puzzlement in looking at the slabs of Stone-henge: our experience is that of being unable to read, knowing, mean-while, that there is a message to be made out. And this is to be experiencing the present in the light of the past.

There are two exercises which are indispensable to the analytical philosopher of history. The first is to try to imagine what it would be like to experience the present without knowing anything at all about the past. The second is to try to imagine what it would be like to experience the present if we *also* knew the future. Concerning the latter, I shall have more to say later on. But concerning the former, this much at least can be said, that our experience of the *present* is very much a matter which depends upon our knowledge of the past. We will, then, necessarily experience the present differently in accordance with the different pasts to which we are able to connect it. Proust is full of instructive examples of this. There is, for example, the inability of Mme Sazarin to experience Mme de Villeparisis as an old woman, for she recalls that woman when she was full of a fatal charm, and from a time when she exercised an immense erotic power over Mme Sazarin's father; while Marcel, who has no access to that past, is unable to experience Mme de Villeparisis as anything other than an old woman in connection with whom erotic considerations can scarcely arise. I emphasize this before turning once more to Beard. It is not simply that present factors tend to distort our statements about the *past*. *Past* factors tend to distort our experience of the present, so that, loosely speaking, one might say that Marcel and Mme Sazarin were hardly experiencing the *same person*. It is, as we shall see, as difficult to extract the past from the present as it is to join the future with the present. The Rondanini Pietà was carved from a Greek column. One could see that column, before its transformation, as ancient. But who could see it in, say, the fourteenth century as the column out of which the Rondanini Pietà was to be carved by Michelangelo?

Beard, and historical relativists generally, complain of the advantages enjoyed by scientists which are unavailable to historians. He says, for example, that unlike the scientist, the historian cannot observe his subject matter: 'He cannot see it *objectively* as the chemist sees his test tubes and compounds.'[1] This is in some way an odd kind of lamentation. Our incapacity, which is granted, to observe the past, is not a defect in history itself, but a deficiency which it is the precise purpose of history to overcome. It is not, comparably, a deficiency in medical science that people fall ill, but rather, the deficiency in human existence in which illness consists is precisely what we have medical science *for:* we would not need it if we were always sound. That cities lie at a spatial distance from one another is not to be regarded as a defect in our systems of transportation: it is a deficiency, if one may style it such, which systems of transportation are precisely designed to overcome. It is just because we do not have direct access to the past that we have history to begin with: history owes its *existence* to this fact: it makes history possible rather than impossible or unnecessary.

But it is too obvious to require labouring that scientists do not as a general rule have access, via direct observation, to their subject-matter. It is precisely because what they often deal with *is* unobservable that they have recourse to elaborate theories and techniques, and what scientists can directly observe may stand in no more intimate a relationship to their subject-matter than what historians can observe—medals and manuscripts and potsherds—stands to theirs. The unobservability of subject-matter may be due to different reasons. It is perhaps a logical truth that we cannot observe the past, and merely a matter of contingent fact that we cannot observe electrons or genes. But that the grounds should be logical in the one case and factual in the other does not entail any difference in current *practice*. There may be branches of science in which the subject-matter is observable: certain parts of chemistry, of zoology, and geology, for instance, come to mind. But these are apt to be very elementary sciences in contrast with, say, atomic physics, and the fact that the most highly developed sciences are concerned specifically with unobservables shows, I think, that the unobservability of subject-matter is not an overwhelming disadvantage for science, or that access to subject matter is an overwhelming advantage. Therefore, when Beard complains that the 'historians must "see" the actuality of history through the medium of documentation. That is his sole recourse'—we need waste little sympathy. It is too common a feature of scientific work for it to raise any *special* problems for history. I have argued that without reference to history-as-actuality, historians could not so much as see documents as such. And comparably, in science, without relying upon concepts concerned with unobservable entities, scientists could not read the marks on photographic plates or the tracks in cloud chambers or oscilloscopes. Again, this is an obvious point. It takes a considerable training in theoretical work before one is in a position to see the relevant aspects of such marks. Optically we may be the scientist's equal, but we do not, for all that, see things as he does.[1] And surely this is the case with historical documents. If we are unable to read these documents, it is not glasses but historiographical instruction which is required. There are plain and obvious differences between history and the sciences, but they do not lie here.

There is a second contrast between history and science which Beard draws, and this brings us to the heart of the relativist's chief position. It is a contrast which turns upon a different sense of the word 'objective', the sense, namely, in which 'objective' is contrasted with 'biased'. There is something inherent in the subject-matter dealt with by scientists and historians respectively which permits an attitude of neutrality on the scientists' part and which forces a partisan attitude on the historians. 'The events and personalities of history in their very nature involve ethical and aesthetic considerations,' Beard writes, while, by contrast, 'events in chemistry and physics [invite] neutrality on the part of the "observer".'[2] It is true that historians deal with human beings, and that human beings have attitudes. But that this entails any difference between historical and scientific practices is surely wrong, and questions of entailment apart, it is remarkable that an historian should make such a claim. The slightest familiarity with the history of science would contradict it. In the seventeenth century, for example, scientists were overwhelmingly motivated by ethical and aesthetic considerations, nor could it easily have been otherwise, given the religious and moral atmosphere in which they worked. It is doubtful if these considerations are wholly absent from science today. The image of the dispassionate and coldly objective scientist is itself an ethical ideal, and even if it is one to which a good many scientists perhaps conform, it would be evidence of the most extraordinary naïveté to suppose that scientists are in fact as neutral as we should like for them to be. There are just too many rewards connected with scientific achievement, not to mention the sheer desire to be *right*, for us to suppose that scientists do not have to make sometimes quite strenuous efforts towards being neutral. Claude Bernard wrote:

Men who have an excessive faith in their own theories or in their ideas are not only poorly disposed to make discoveries, but they also make very poor observations. They necessarily observe with a preconceived idea, and, when they have begun an experiment, they want to see in its results only a confirmation of their theory. . . . It quite naturally happens that those who believe too much in their own theories do not sufficiently believe in the theories of others. . . . The conclusion of all this is that it is necessary to obliterate ones opinion as well as that of others when faced with the decisions of the experiment.[1]

Pierre Duhem comments on this passage:

Such a rule is not easily followed. It requires of the scientist an absolute detachment from his own thought and a complete absense of animosity when confronted with the opinion of another person; neither vanity nor envy ought to be countenanced by him. . . . Freedom of mind, which constitutes the sole principle of experimental method, according to Claude Bernard, does not depend merely on intellectual conditions, but also on moral conditions, making its practice rarer and more meritorious.[2]

Note that the deliberate distortion of results is not what is in issue here. Beard does not mean of historians, nor Bernard and Duhem of scientists, that it is hard for them to forebear from forgery and hoax. It is rather a case of *involuntary* distortion, induced for whatever reason. But here again, Duhem's observation that it is *hard* to retain objectivity; that retaining it is altogether praiseworthy and hence not automatic on the part of a scientist, goes a considerable distance towards softening, if not, indeed, obliterating Beard's distinction. A disposition towards bias is something common to the different disciplines.

There are certain factors which, in Beard's view at least, finally render nugatory any attempt by historians to maintain an openness to fact and a freedom of mind: 'Whatever acts of purification the historian may perform, he remains human, a creature of time, place, circumstance, interest, and culture.'[3] It is hard to see whether there is here an implicit slur or an implicit compliment for *scientists*. But if the fact that they too are human, and have human interests and predilictions, is not incompatible with their finding truth, it is difficult to see why it should be different for historians. Historians from different backgrounds may naturally be concerned with different things, but their differing provenances surely do not automatically require that they would not each be making true statements. The classical philosophical criticism of the relativist view is that it rests upon a confusion of the causes of a belief and the reasons for a belief. The fact that there should be *causes* for a belief is utterly independent of the question as to whether that belief is well grounded, and *that*

7   DAPH

question we can decide in utter ignorance of the causes which may have operated on the man who held it. A writer may, because of his Catholic background, take a favourable view of the proposition that the Borgia family were virtuous to a member, and that they have been badly maligned. It is up to him to make his case, however, and it is on the basis of the evidence he turns up that we decide for him or against him. Indeed his motives may make him sensitive to documents and possible interpretations of them which *less* committed historians might never have noticed. Scientists, no less than historians, are subject to causes. But in science, no more than in history, are beliefs accepted or rejected on *ad hominem* grounds. There are few more pernicious beliefs than the one which suggests that we have cast serious doubts upon a belief by explaining why someone came to hold it. One might, for instance, explain Beard's belief by appeal to the fact that he was an historian, and hence much concerned with the *causes* of mens' beliefs. But to *point out* that fact does not *refute* Beard's thesis, nor can we say that his thesis is erroneous *because he was an historian*. It is erroneous in the way in which any thesis which fails to square with the facts is. This does not mean that what Beard says is not important however. I myself have been insisting that the time at which they are made is one of the factors which have to be taken into consideration when we evaluate an historical sentence, and it may very well be true that this is not always the case with scientific sentences when the latter, for instance, make no specific temporal claims.

Even here, of course, some careful qualification must be made, for it has been clearly established that in some sense it is important to assign a date to a given theory. For instance, Ernest Nagel has shown that we cannot,[1] without some such restriction, speak of the reducibility of thermodynamics to mechanics. If we are thinking of the mechanics of Newton's time, the reduction will not go through. So we have to speak of the mechanical theories as held at a certain date. Nevertheless, it is a contingent fact that one mechanical theory should have been held in the seventeenth century, and another in the nineteenth century. In regard to the logic of reduction, no essentially temporal reference is required, and this is not quite the case with history. But this is a matter to which I shall turn later. I want now to bring out what I think is the essential misconception in Beard's analysis.

There are many crucial differences between history and science, but, in our first skirmishes with Beard, we have been able, I think, to see that he has almost uncannily managed not to identify any of them. I shall endeavour to explain this curious blindness by suggesting that Beard had an almost total misconception of science. Holding, as he did, to an erroneous picture of science, he contrasted it with history, a discipline he practised with paramount skill. Finding a discrepancy between history, as he understood it, and science, as he misunderstood it, he concluded that there were inherent defects in history which had nothing corresponding to them in the sciences. These 'defects' are, in fact, inherent features of science, and once we have come to recognize this, we will be able at once to see the source of Beard's error, and rid ourselves of the contrast between history and science as he framed it. But it is not merely that these so-called defects in history are something to be found in science as well. They are rather *sine qua non* for *empirical* inquiry, *including* history.

What Beard found unsettling about history was not so much that historians' hypotheses are *caused*, but that historians must have recourse to hypotheses at all. As though the use of hypotheses were some fatally disfiguring *faute de mieux* in historical inquiry, brought in as a consequence of our granted incapacity to see and observe what it is that we are interested in finding out about. It is precisely here that we can begin to discern the misleadingness of the analogy, or metaphor, that we 'see' the past (history-as-actuality) through the medium of documents (history-as-record). The prejudice here is that we know what we can see, and only what we can see, and that, accordingly, if we are to know the past, we must somehow be able to see it: for otherwise (and this was the problem that tormented Lewis) how can we know the past at all? While there can be little doubt that observing is an essential, indeed uneliminable feature of empirical knowledge, it is not by any means the whole of it. It is sufficiently striking a feature of scientific inquiry, however, that, coupling it with the prejudice which identifies knowledge with observation, one can understand the great desire of empiricism to effect a total translation of the language of science into observational vocabulary. Hence the appeal of Pragmatism and Phenomenalism as variants of the radical empiricist programme, an appeal dimly sensed by Beard, who wrote as one who believed that major premises are somehow fatally disfiguring blemishes in syllogisms, or that, at any rate, the use of syllogisms is, in some sense, a fatally disfiguring blemish in historical work. One has access to the past only through inference, and to make such inferences requires, or presupposes, certain theoretical sentences, whether made explicit or not, which connect present evidence with past fact. But it is not in fact such sentences which primarily concern Beard. It is rather, I think, the fact that we use theories in some manner or other to *organize* the events, evidence for which is to be found in history-as-record. And it was Beard's odd persuasion that theories in this sense are not used in science, that in science one can just see things the way they are. If we could just see the events which concern us as historians, that would be ideal. But we cannot. And so we resort to hypotheses. And this, somehow, is a bad thing.

Beard writes:

Any overarching hypothesis or conception employed to give coherence and structure to past events in written history is an interpretation of some kind, something transcendent.[1]

Now I think it true that historians do, that written history must, employ such 'overarching hypotheses', and I shall later argue this in detail. Here, however, I am concerned to raise two questions. The first is whether this fact about history serves, as such, to make the contrast Beard wishes to establish between history and science. And secondly, whether there might be some contrast *within* history, between pieces of written history which do and which do not make use of such overarching hypotheses. That is to say, even if in fact every piece of written history should prove to have employed these conceptions, could we, at least ideally, conceive of a piece of written history which did not, and which then would satisfy Beard as being the sort of *objective* history which history, to its detriment, has not so far succeeded in achieving? What would such an account look like?

Let us first state, what the argument entitles us to state: that (*a*) a theory may be correct or incorrect independently of what caused somebody to entertain it, and (*b*) as a general rule we determine whether a theory is correct by making observations, and, finally, (*c*) nothing is an observation apart from a theory of some sort. Let us abandon, then, the image of the historian peering through a screen of documents and trying to make out a landscape of past events. Let us rather see him as trying to test, or support, or check up on some *account* of past events by peering at 'history-as-record'. This would give some further sense to my earlier observation that we could not see what we see as history-as-record without implicit reference to history-as-actuality. One wants to find out whether a certain account of the past is correct. One then makes what might properly be called an historical observation. One checks the records, loosely speaking. But notice in passing that 'is a record' is a relational predicate. We speak of things as records *of*. So to be correctly designated a record, the thing so called must already stand in a certain relationship with something else. My point, however, is that one does not go naked into the archives. But then, it might be argued, neither does one go naked into the laboratory. 'It is impossible,' Duhem wrote, 'to leave outside the laboratory door the theory that we wish to test, for without theory it is impossible to regulate a single instrument or interpret a single reading.'[1] Without theories, 'the readings would be devoid of meanings'. It will be obvious that my sympathies are wholly with this account of Duhem's, an account to which, were this a work in philosophy of science, one could devote a great deal of careful analysis. I cite it here, however, because of the marked contrast in which it stands to the implicit picture of science Beard seems to have been dominated by.

Beard, I want to suggest, was essentially a Baconian in philosophy, and there is an essentially Baconian error in his thought. Bacon correctly and importantly indicated that human beings are subject to the distorting influence of a variety of different prejudices which he labelled 'Idols of the Human Mind'. We must, he felt, if we are to advance in knowledge, rid ourselves of these idols, and approach Nature fresh and without preconception. But then, and almost as though he were seduced by a pun, Bacon went from this salutary advice to the decidedly unsalutary injunction that we go to Nature without *theories*. He thought, indeed, that his own method of 'true induction', employed by men of open minds, would permit them, by means of a complex set of tabulations, finally, and without theories, to arrive at the 'form' of phenomena. We know now that Bacon's system is inherently impossible,[2] that science, were it to have heeded Bacon,[3] would have ground to a dead halt, and that Bacon, accordingly, was not the instigator of the scientific revolution he hoped for, since *that* revolution was based upon the use of a method, the hypothetic-deductive method, which Bacon himself would have found repugnant. Beard was a Baconian in the sense that he assumed, not merely that science can, but that science does get on without hypotheses; and that hypotheses are of a piece with the Idols of the Human Mind.

The question 'Can physics be objective ?' sounds strange, I think, but not nearly so strange as *this* reason for a negative answer: physics is not objective because it employs theories and hypotheses. The strangeness of this answer lies in the fact that it is hard to know what, apart from a set of theories and hypotheses, together with testing and evaluating them, one can *mean* by physics anyway. But moreover, while it is a fact that physicists may be objective in varying degrees, may or may not allow personal and philosophical considerations to determine the readings they give of phenomena, this fact would come to very little if, simply by virtue of using *theories*, they were automatically to be disqualified as objective. Were a physicist to take seriously the idea that there is some incompatability between his desire for objectivity and his use of theories, it is not easy to see how, except by finding employment elsewhere, he could continue to regard himself as an objective person. He had better not seek employment as an historian, however. Not if Beard is right. And I have argued that he is right about history. It is only that he is wrong about science, and, for that matter, wrong about the whole structure of empirical knowledge. For the precise criterion he sought to use to distinguish invidiously and despairingly between history and science turns out to be one of the important features they share. The remarkable thing is, however, that he should have been right about history at all. For it is not a common thing to say that history employs conceptions and theories. Usually, indeed, just the *reverse* contrast is drawn between history and science, and the two are opposed in roughly the way in which fact and theory are.

At this point I should like to consider a variant of Beard's views which has been defended by Professor W. H. Walsh, a variant whose examination will help, I hope, to show some further philosophical features of the Relativist's position. Walsh, to be sure, is highly critical of Beard's theses, but nevertheless feels that something can, in the end, be said in their favour. He points out that historians often have theoretical commitments of rather a special sort, and that, though some historians, for example Marxists, exhibit these more dramatically than others, some such set of commitments is adhered to by all historians. There are schemes of interpretation he suggests, which an historian is likely to insist upon even in the face of what might appear to others as overwhelmingly contrary evidence. And some theories are held 'with greater confidence than . . . if they were merely empirical hypotheses'.[1] Assuming there are such theories, what would account for the pertinacity with which they are asserted? Walsh finds the answer to be 'differing philosophical considerations . . . moral and metaphysical beliefs . . . a general philosophy which is confirmed in many fields'.[2] Historians, then, 'approach the past each with his own philosophical ideas, and that . . . has a decisive effect on the way they interpret it.' But then, he contends, it is difficult to see how historians do otherwise: we could not 'even begin to understand unless we presupposed some propositions about human nature, unless we applied some notion of what is reasonable or normal in human behaviour'.[3] If this is so, then the historian's interpretations of the past will in some measure depend upon that set of presuppositions concerning human behaviour he is committed to; and historians, with differing philosophical commitments, will give interpretations which differ from these. Walsh goes on to say that this line of reasoning is simply an extension to human behaviour of that followed by Hume in his famous repudiation of miracles. Hume maintained that 'we cannot give credence to accounts in the past the occurrence of which would have abrogated the laws of physical nature'.[4] It was because he was philosophically committed to certain presuppositions regarding what is 'normal or reasonable' in nature that Hume would have given a considerably different interpretation of biblical events than someone who was, say, more pious or more accommodating.

Here again there is no point in denying what is plain fact. Historians do, with varying degrees of explicitness, adhere to often varying sets of presuppositions regarding the manner in which human beings behave. Let us grant even that they must, if dealing with human behaviour, adhere to *some* such set of preconceptions. The question for the moment is whether these are facts of a kind uniquely pertinent to history, and it seems obvious to me that they are not. We have in any field of inquiry some criteria concerning normal behaviour, and tend to be hostile to or suspicious of accounts which seem to violate these criteria. And we are often inclined to insist upon our criteria here for rather a long time before we will so much as admit that we have genuinely contravening evidence. But this is so in science no less than in history. Considering just Hume's argument, the alleged occurrence of miracles would be at least as incompatible with our criteria for accepted *physical* behaviour as they would be with our criteria for human behaviour. And indeed it was as a consequence of an apparent conflict between miraculous claims, on the one hand, and physical theories on the other, that Hume felt rationally constrained to reject the former.

The matter does not quite end here, however, and what Beard and Walsh together are insisting upon may yet bring out some important fact about history. Someone might contend that we can easily enough tell when this historian or that is a committed Marxist or a committed Freudian. We can readily identify the theoretical presuppositions in such cases. But Walsh wishes to emphasize that it is not merely those historians who are unique in having sets of preconceptions. We are not without them ourselves, only ours are perhaps different, and if we reject their preconceptions, it is only because we are committed to a set of our own, a set perhaps so deeply embedded in our general conceptual scheme that we are hardly conscious of the fact that we have them. Indeed, what passes for 'common sense' might be precisely just such a set of preconceptions. It is perhaps not easy to state these preconceptions in any exhaustively explicit way. Nevertheless, we might roughly be able to say at any given moment when something fails to square with what we regard as normal or reasonable behaviour. We feel a certain shock or register a certain feeling of surprise upon hearing of some piece of behaviour, and this shock or surprise is a sign that our common-sense views have been offended. If the shock is great enough we are disposed simply to reject the account as incredible. This must surely be what Walsh has in mind when he writes as follows:

I cannot escape, if I am to make any sense of my material, making some general judgements about human nature, and in these I shall find my own views constantly cropping up. I shall find myself involuntarily shocked by this event and pleased by that, unconsciously seeing this action as reasonable and that as the reverse.[1]

Now so stated, this is, I think, a very much milder view of things than Hume's. For there is, here, no suggestion that we are to employ our sense of shock as a criterion for what must actually have happened or must not have happened. It is hardly incompatible with common sense to suppose that strange and conceptually shocking things do in fact occur, much less queer and morally shocking things. I may find Suetonious's account of the *dolce vita* of the Emperor Nero morally shocking, and you may take it in your stride: I may be pleased that he is done in, and you may blandly accept the fact. You share neither my indignation nor my jubilation. But this need be the only disagreement between us. Indeed, unless we agreed on the facts, there would be no room for a genuine moral difference. That we morally differ is hardly incompatible with our mutual capacity to regard the same accounts as true or false. So the factors which Walsh here draws to our attention may affect our moral attitudes towards events, but leave quite unaffected our powers as historians. Surely he must have meant something more exciting than this if he is to couple his own views with those of Hume. So I shall assume that what he meant to say is this. Granted that common sense permits the occurrence of some surprising and shocking things, relative to a given set of presuppositions—and there would be no such thing as a shock or surprise if we *didn't* have some preconceptions[2]—there are certain accounts which we simply would not *allow* as true no matter what the 'evidence' for them might be. For instance, if someone were to say that Plato wrote the *Laws* when he was three days old, using 'three days old' as it is in fact used, and designating by the *Laws* just that piece of writing customarily so designated, we would, I think, generally consider that no amount of evidence could get us to believe this statement.[3] So let us say that there are certain preconceptions, such that any account which is incompatible with them is rejected as historically inadmissible. This was more or less Hume's position on miracles. And it was more or less Bradley's point, as Walsh himself suggests, when, in the *Presuppositions of a Critical History*, Bradley proposes that we are to regard as believable (or admissible) only those accounts of events which have some analogy in present experience.[4]

Criteria for the admissibility of historical accounts carry us, of course, very scant distance in historical work. It tells us at best which accounts are believable, but there is no compulsion to *believe* a believable account. That the philosopher Kant took his mistress to Crete is *believable*. It even has analogues in present experience: some people take their mistresses to Crete. For all that, we are not required to *believe* the statement, but only to allow it is not ruled out by our preconceptions. Bradley was looking to find a criterion for ruling out certain accounts as *impossible*. Accounts of miracles are not, in fact, in violation of any logical criteria: they are, commonly, *logically* possible. But Bradley wanted a notion of empirical impossibility, and recommended that we use consonance with present experience as a criterion for empirical possibility. But not every *possible* account is *true*, nor, if I may be pardoned for so speaking, is every possible event actual.

But it is hardly to be expected that historians should, in fact, differ very markedly over their criteria for *impossible* accounts. Indeed, we might very nearly appeal to Bradley's own criterion by way of ruling out the possibility of there having ever been historians whose criteria of possibility were significantly different from ours: for we could find nothing analogous to such a person in *present* experience. And we might almost appeal to Hume's criterion of veracity in such cases, and ask whether the probability is not higher that such a personage was writing fiction than that the events he narrates should have ever happened. It would almost be the case that we would not allow, by our criteria, that such a person be considered an historian, whatever 'evidence' he might produce that he was one. Nevertheless, if this is what it comes down to, we have not achieved much. Historians with shared criteria of admissibility may give conflicting accounts, and the fact that we share these criteria would only permit us to say that both accounts are believable. For if one account is possible, it hardly follows that the other is *impossible*. *Each* of a pair of conflicting accounts may have some analogue in present experience.

We are, nevertheless, now in a position to state the form of historical relativism which these considerations suggest. It is that certain accounts are to be construed as *possible* relative to a set of preconceptions, and that any account inconsistent with these preconceptions will not be tolerated by historians whose preconceptions they are. But there may be differing sets of preconceptions. Accordingly, a given account A may be possible relative to one set and impossible relative to another. Notice, incidentally, that if A is impossible relative to a set of presuppositions P, one would *not*, as it were, be insisting upon P in the face of contrary evidence, that is, evidence in favour of A. If A is impossible in the light of P, there just could not be evidence for it, so far as those who subscribe to P are concerned. One could only admit evidence for accounts which were *possible* relative to one's presuppositions. This would create something of a problem so far as accounting for changes in peoples' preconceptions was concerned, but it may very well be that these are not changed as a consequence of evidence coming to tell against them. At all events, I should like to examine briefly this version of relativism, a version, I think, which is invariant to the distinction between history and science.

Suppose we take an ideal case where two historians subscribe to a pair of conflicting historical sentences S-1 and S-2. Suppose, moreover, that the first historian has a set of preconceptions P-1, relative to which S-1 is admissible, and S-2 not; while the second historian subscribes to a set of preconceptions P-2, relative to which S-2 is admissible and S-1 not. It is important that we recognize that 'admissible' and 'impossible' are not, as it were, simple properties of sentences: we had better speak of 'admissible relative to' and 'impossible relative to' where what it is that statements are admissible or impossible relative to are sets of preconceptions. But let us emphasize the consequences of this relativization.

The first is that once we have relativized the sentences at issue, there can be no logical opposition between the results. Even though S-1 and S-2 are in logical opposition, 'S-1 is acceptable relative to P-1' and 'S-2 is acceptable relative to P-2' are not merely not logically opposed: they are both true. This result is similar to what would take place if one were to insist that every statement be relativized to its utterer. In that case, two sentences p and q, originally in opposition, become, once relativized, perfectly compatible. That is 'A says that p' and 'B says that q' are not merely not incompatible: they are both true.

But secondly, there can be no genuine disagreement between historians regarding the acceptability or impossibility of historical sentences. For either they share the *same* set of presuppositions, and hence cannot genuinely disagree, or they have different sets of presuppositions, in which case they cannot genuinely disagree. They cannot genuinely disagree here because they must agree that a mooted sentence is acceptable relative to the presuppositions of the one and impossible relative to the presupposition of the other. To be sure, one might go on to say that there can nevertheless be some genuine conflict between sets of *presuppositions*. Or one can say this unless some further relativization occurs, for example they might have different criteria for the acceptability of a set of presuppositions, in which case there is no genuine disagreement about presuppositions, but at best a disagreement about criteria. My point is that there can be genuine disagreements only where there is common ground, and only where one may bring things into logical opposition. Otherwise we continue to dissolve disagreements through relativization.

With this we can finally evaluate one of the chief arguments sometimes brought forward in defence of historical relativism. Walsh writes: 'there is without doubt a *prima facie* case for an ultimate historical scepticism, a case which the spectacle of actual differences amongst historians greatly strengthens.'[1] But in fact the case is weakened if these differences amongst historians are genuine, and then only if they are differences of the right *kind*. For there are, I submit, levels of disagreement. To make the matter simple, let us suppose that there are just three levels of disagreement: disagreement over historical statements, over presuppositions, and over criteria for a given set of presuppositions. Now it seems to me that the bulk of historians' differences arise at the first level, over statements which are equally *acceptable*, the question being only over which of them, if either, is true. Historians, for example, differ over the question whether Caesar was or was not in Britain. But the fact that they differ lends not a scrap of support to the case for historical scepticism, for surely the statement that he was in Britain is acceptable to most historians, and impossible to very few. But the fact that it is acceptable to most entails that all historians to whom it is acceptable share the same set of presuppositions: otherwise none of them could genuinely differ. So there is a whole class of differences we may automatically discount.

Nevertheless, there can be disagreements over presuppositions, for instance, presuppositions regarding normal or reasonable human behaviour, to take Walsh's example. Such disagreements surely exist. Nevertheless, each of a set of differing sets of presuppositions may, in turn, be equally acceptable to historians who share criteria for the acceptability of sets of presuppositions. It does not follow from the fact that a theory is acceptable that people will in fact accept it, for acceptability entails, not that the theory is correct, but that it satisfies the criteria for a theory. The phlogiston theory, for example, is scientifically acceptable, but nobody any longer accepts it. Indeed, it is precisely as a consequence of its scientific acceptability that it is not any longer accepted: had it failed to satisfy the criteria for a scientific theory, it would have been rejected for far different sorts of reasons than it was in fact rejected: if it failed to satisfy those criteria, it would not even have been regarded as a theory. So, in general, differences over presuppositions lend not a scrap of support to historical scepticism. For such differences are removable in principle, so long as those who differ in fact share criteria of acceptability for sets of presuppositions. And there can be little doubt that we change our presuppositions regarding rational human actions from time to time. So, once more, we can discount a whole class of disagreements as either irrelevant or inimical to historical relativism.

But, finally, there may be differences of a more ultimate sort. Men may differ on the very criteria in accordance with which they would adjudicate between theories, even general philosophical theories. The so-called conflict between science and religion might very well be an instance of this ultimate sort of disagreement, a disagreement so deep that there is nothing those who differ may appeal to as common ground. And it is just here, I think, that we can make a connection with the sorts of *moral* considerations which Walsh alluded to. There are perhaps differences of belief which resemble differences of attitude, differences of so fundamental a sort that we might term them *disagreements of principle*. One can perhaps do little better here than to speak of certain basic decisions, decisions of a sort which will determine what other kinds of decisions one is to make at higher levels. And such differences may disappear only when one or another party decides to cross over to the other side, to change his ultimate commitments. So, in a general way, one might say that whatever we believe finally is relative to some such basic decision, and that, in an important sense, such decisions are arbitrary. They are arbitrary in the sense that they are not made in accordance with any criteria, for they determine, finally, what are to be the criteria we shall accept. But having allowed this much, we can, I think, now bring this entire discussion to an end by emphasizing that if one chooses to regard these facts as a basis for a scepticism with regard to history, one has no good grounds for doing so. Not because history is not relative to such basic decisions, but because every human cognitive enterprise is. One could not be sceptical about history without being sceptical about everything else, and this, finally, destroys whatever specific force relativism might be thought to have with regard to history. It was as though a man were to lament that it is a sad thing to be a Frenchman, for all Frenchmen die. He may easily and obviously be disabused of his melancholy through having it pointed out to him that Frenchmen are not uniquely mortal. If he persists, saying that he knows this, but what a pity it is that *Frenchmen* should die, we can point out that there is no reason why Frenchmen should be specially privileged, unless one has some peculiar prejudice. And so it is with history. History is no more and no less subject to the relativistic factors than science is. And if one says that there is a special pity in the fact that history is, there is nothing to say except this is a prejudice, and that one could not legitimately demand any exemptions here.

Walsh's argument, then, comes to very little because it comes to rather too much. If one were to demonstrate that it is impossible to make a true statement, it would follow, of course, that one could not make a true statement about the past. But why then call this *historical* scepticism? It is scepticism '*überhaubt*,' and we are not obliged to deal with '*überhaubt*' scepticism.

There were, however, two contrasts in Beard I wished to discuss. The first was his contrast between history and science. This turned out to be illegitimate. It was based upon a total misconception of science in that it suggested that science does not, while history does (to history's detriment), employ certain overarching schemes of organization which go beyond what is given. We destroyed this contrast through pointing out that the employment of such organizational schemes was a generic feature of empirical knowledge. The second contrast was within history itself, a contrast between history which employs such schemes and history which does not. The question is whether, even ideally, there can be history of the latter sort. I shall now proceed to argue that this contrast too is bogus. To be sure, this might be said to follow from our results so far, so that no further argument here is required. Nevertheless, the matter demands some special analysis, and in carrying this out I shall be interested in making two points. First of all, there is an essential mistake, though an understandable one, in the model of historical activity implicit in Beard's language: that *there* is history-as-actuality, and *here* is history-as-record, and that it is the task of the historian to seek to reproduce (via history-as-thought) the former by means of the latter, though never quite succeeding. I shall try to show that we cannot succeed in this for rather different kinds of reasons than mere paucity of documentation, and I shall try to bring this out by trying to imagine what a *perfect* account would look like. Having seen why we cannot have a perfect account, we shall, I hope, see why it is not even an ideal for history to achieve, and that *in the nature of the case* historians are obliged to aim, not at a reproduction but at a kind of organization of the past. And this, finally, I shall try to exhibit as logically dependent upon topical interests which motivate historians, so that, if I am right, historical relativism will finally be vindicated. It will be vindicated in the sense that it is, in a general way, correct, and that we cannot conceive of history without organizational schemes, nor of historically organizing schemes apart from specific human interests.

My second point will be this. The difference between history and science is not that history does and science does not employ organizing schemes which go beyond what is given. Both do. The difference has to do with the *kind* of organizing schemes employed by each. *History* tells stories.

VII

HISTORY AND CHRONICLE

I began by saying that historians seek to make true statements about their past. And I have been maintaining, against certain philosophical arguments to the contrary, that they can in principle succeed in doing this, so the question, if I have been right, is not whether they can, but whether they do succeed in making such statements. That they do succeed in this I do not doubt, but I now wish to consider what further can be said regarding the kinds of statements it is their purpose to establish. Often, I think, the statements they make may be regarded as explicit answers to what I have elsewhere called 'historical questions';[1] questions of the form 'What happened at x?' where 'x' stands for a spatial region during some past interval of time. The answers, even to the same historical question, may have varying degrees of explicitness and detail. Asked, for instance, what happened at Waterloo in 1815, I may answer merely 'Napoleon lost'. And this may be a perfectly good answer if this is all that he who posed the question wished to know. For people enter into historical questioning with varying amounts of antecedent information. We may indeed say that whole books exist which answer just the same question that 'Napoleon lost' also answers. Let us then say that one can specify a range of statements, relative to a single historical question, which differ in point of detail. These statements will all be about the same event, for example the Battle of Waterloo, but tell increasing numbers of things about it. At the opposite end of the range from the bland statement 'Napoleon lost' is what we might term the *maximally detailed account* of the Battle of Waterloo. It is *that* end of the range which will occupy me now.

It is sometimes contended that it is not merely the aim of historians to make true statements about the past, but to give, ideally, the maximally detailed statement about the past. And the question I shall concern myself with is whether they can, at least in principle, succeed at this end of the range, given that we have been correct in concluding that they can succeed with statements at the other end.

Beard, at one point, complains that we cannot, though this, as a general rule, is a consequence of the fact that there are always parts of history-as-actuality for which we have no history-as-record, or at least none that we know about.[1] So, given that there are gaps in history-as-record, there are corresponding gaps in history-as-thought, holes, as it were, in our knowledge of the past. So *in fact* we have something which is always less than perfect knowledge of history-as-actuality, and sometimes, indeed, Beard means, by historical relativism, that our knowledge of the past is relative to a body of evidence actually in the possession of historians. But I want to know whether it is *only* a matter of fact that we have less than perfect knowledge here. And this can hardly be answered until we have some rather clearer idea than I think we have of what perfect knowledge would consist in. This, however, is a question which is not merely capable of being raised with regard to things past, things which, because they are past, are incapable of being known directly and must be found out about on the basis of what we can observe. We might raise it about things that we can in fact observe. With such things, I suppose, there is no problem of evidence, for we have, or can have, the thing itself to scrutinize. If there is a difficulty, it might lie elsewhere, and specifically, I think, it lies in the question of giving some meaning to the expression 'perfect knowledge of *x*'. What, for instance, would it be like to have perfect knowledge of the Empire State Building? Or the apple on the table? Or Brigitte Bardot? And if we have difficulty in saying what we mean by 'perfect knowledge' of contemporary objects, the lament that we lack perfect knowledge of the past is not very impressive. For the problem would have nothing to do with presentness or pastness, but with the very notion of perfectly knowing something.

The lament is nevertheless an interesting symptom in the syndrome of relativism, for it helps us to see why Beard was so deprecating of his profession. It was not merely that he saw it as defective in comparison with science, but defective, as well, in terms of its own implicit ideal of achieving the perfect account of events for which we have at best imperfect ones. And he blamed history-as-record for this instead of questioning the validity of the ideal itself. Imagine an artist who subscribed to the Imitation Theory of art, and who became so obsessed with the imitation of reality, which falls always short of reproducing the subject, that he decided only the thing itself will do as an imitation of itself. He tries, accordingly, to go *all the way*, duplicating the landscape, using real trees, real water, real birds. Perfect success would, of course, be utter failure. For he would have produced, as a consequence of all his labours, not a work of art but the subject for one, and the labour of now making paintings of it would remain to be done. Not being what it is a picture of is not a defect in pictures, but a necessary condition for something to be a picture at all. And it is a mistake to suppose that everything which is in the subject must be reproduced in the picture when it is quite enough only that whatever is in the picture should be in the subject, or correspond to something there. Pictures, in the nature of the case, leave things out. And we may say as much of histories of things. What Beard fails to understand is that even if we could witness the whole past, any account we would give of it would involve selection, emphasis, elimination, and would presuppose some criteria of relevance, so that our account could not, unless it wished to fail through succeeding, include everything. It is true, of course, that there are gaps in the record, questions we would like to have answers to which we cannot have because we lack data. But this incontrovertible fact only conceals Beard's real complaint. Comparably, a man might want desperately to paint, but cannot because it is raining outside or the paint-shop is closed. But it would be almost pointless to mention these facts in connection with an artist who thinks of painting as actually duplicating his subjects. His incapacities are logical and not contingent, for he does not want to do art, he wants to be God, and painting for him is an unsatisfactory *faute de mieux*. This is an old platonic attitude, which considered it a scandal that pictures of beds were not themselves real beds, much less Real Beds. Beard's Baconian attitude towards science, which lay at the heart of one of the illegitimate contrasts he drew, was complicated by a platonic attitude towards art, or history, and this lay at the heart of the other illegitimate contrast. History-as-thought is the faulty imitation of history-as-actuality, where the term *faulty* is not used, as it were, to distinguish amongst imitations, but to characterize imitations as a class: an imitation of *x* just is not *x*. Something is missing. So we do not have perfect accounts, though this in part is due to the fact that an account of *x* is not *x*, and indeed can only be an account of *x* if there are things about *x* it leaves out.

I shall, however, return to the notion of a perfect account later. For I wish here to introduce another and different view of the task of historians. It is a view which, in a way, accepts the ideal of imitation of the past, but wants to insist that there is something beyond giving accounts, even perfect accounts, of the past, or parts of the past, which it is also the aim of history to do. For in addition to making true statements about the past, it is held, historians are interested in giving *interpretations* of the past. And even if we had a perfect account, the task of interpretation would remain to be done. The problem of just giving descriptions belongs to a humbler level of historical work: it is, indeed, the work of chroniclers. This is a distinction I am unable to accept. For I wish to maintain that history is all of a piece. It is all of a piece in the sense that there is nothing one might call a pure description in contrast with something else to be called an interpretation. Just to do history at all is to employ some over-arching conception which, in Beard's terms, go beyond what is given.[1] And to see that this is so is to see that history as an imitation or duplication of the past is an impossible ideal. Once these points have been established, we may, I think, go back to the notion of a perfect account somewhat better prepared to understand what, finally, is wrong with that notion. And I shall try to show that the reasons why we are unable to give a perfect account of the past do not have so much to do with the concept of an account, or to any fact about the past, or to gaps in history-as-record, but rather, and far more importantly, to certain facts about the future. Indeed, I shall contend, what *ultimately* makes the perfect account unfeasible is precisely what makes speculative philosophy of history unfeasible. I shall, accordingly, be engaged in a somewhat complicated polemic, but I shall begin with defending the notion that history is all of a piece. This will take up the entirety of this chapter, for there is a great deal involved in the issues it concerns.

The distinction between history and chronicle or, more invidiously, between *mere* chronicle and history *proper*, is frequently to be encountered in philosophical writings on history, and it is made with various intents.

Croce, for example, made the distinction with respect to accounts of those parts of the past in which we are vitally interested, in contrast with accounts that connect with no such vital interests, the latter being chronicles.[1] Chronicle, then, is academic history, though Croce trivialized his point by suggesting that we never at any rate write the history of things we are not interested in, so that *all* history, to use his celebrated *mot*, is contemporary history. In that case we could not write chronicles if we wanted to. Croce, to be sure, is infuriatingly inconstant in his writing, and it is hard to attach any single meaning to his celebrated slogan. He sometimes means, not that a piece of history must answer to some present interest, but rather must report something which has only an analogue in present experience, and, if it has no analogue, *then* it is chronicle and not history: he never, after all, said that all *chronicle* is 'present' chronicle. But apart from connecting with the present in one or another way, there is for him no further difference between history and chronicle, and certainly no formal difference. Chronicle, as he puts it, is 'dead history' while history is 'live chronicle', which is a little like saying that a man is a live corpse while a corpse is a dead man. Whatever the case, it is not this form of the distinction which is important for us.

A more pertinent way of framing the distinction is this. Chronicle is said to be just an account of what happened, and nothing more than that. It is a statement, of whatever degree of complexity, which lies in the range one end of which is given by the 'perfect account'. In fact, the perfect account, were it possible to state it, would still be nothing more than chronicle, for it would differ from other statements in the range only quantitatively, giving more details. Giving, indeed, *all* the details. So the very best kind of chronicle would still not quite be history in the proper sense, and something could then be properly a piece of history even if it reported far fewer details than the perfect account. *Proper* history regards chronicles as preparatory exercises. Its *own* task is rather concerned with assigning some meaning to, or discerning some meaning in, the facts allegedly reported by chronicles. Some such view as this seems to have been held by Professor Walsh, who sees two possibilities for historical work:

The first is that the historian confines himself (or should confine himself) to an exact description of what happened, constructing what might be called a plain narrative of past events. The other is that he goes beyond such a plain narrative, and aims not merely at saying what happened, but also (in some sense) explaining it. In the second case the kind of narrative he constructs may be described as 'significant' rather than 'plain'.[1]

Chronicles, then, would be plain narratives; and history proper would be expressed in significant narratives. This is just the view I want to examine.

Let me first of all suggest that whatever piece of historical writing one chooses as an instance of a chronicle, or as even closely approximating a chronicle, it must do *something more* than satisfy the following two necessary conditions for any piece of history: *any* piece of history must (*a*) report events which actually happened; and (*b*) report them in the order of their occurrence, or, rather, enable us to tell in what order the events did occur.

I take it that these necessary conditions are not controversial, that they state the very least we expect of a piece of history, even though they by no means constitute, jointly, a *sufficient* condition for something to qualify as a piece of history. This is readily demonstrated. For anyone can produce something which satisfies (*a*) and (*b*) and which would *not* be admitted to be a piece of written history. Here, for instance, is one.

S: Naram-Sin built the Sun Temple at Sippar; then Phillip III exiled the Moriscos; then Urguiza defeated the forces of Bueno Aires at Cepada; then Arthur Danto awoke on the stroke of seven, 20 October 1961.

Not merely is *S* not a narrative, but I think it readily demonstrable that a *significant* narrative is not merely a statement which satifies (*a*) and (*b*) together with a further necessary condition: (*c*) it explains what happened.

That a statement which satisfies all three of these conditions can still fall short of being a significant narrative is plainly demonstrable, for one can readily produce a statement which satisfies these conditions and is not a narrative *at all*. Here, for instance, is one.

S′:   Naram-Sin built the Sun Temple at Sippar as a consequence of pressure brought on him by the priestly class; then Phillip III exiled the Moriscos because of his religious convictions; then Urguiza defeated the forces of Buenos Aires at Cepada because he was better equipped; then Arthur Danto woke on the stroke of seven, 20 October 1961, because he wanted to get an early start for the excavations at Cervetri.

Accordingly, (*a*) through (*c*) cannot be said to constitute a sufficient condition for a significant narrative.

It may be objected here that these examples are not fair, since what was intended was a distinction between narratives, and neither *S* nor *S'* is one of these. This is a just criticism. But I have at least shown that '*N* is a narrative' cannot be analysed as '*N* satisfies (*a*) through (*c*)', much less '*N* satisfies (*a*) and (*b*)'; and for the moment this is enough. If we consider it to be the aim of historians to write narratives, then they plainly must do something more than describe things that happened, in the order in which they happened, even if, in addition, they explain why the things they describe did happen, and *even* if they correctly explain them, as *S'* does not. (What is wrong with *S'* is *not* that its explanations happen to be incorrect.) Whatever this something more is to be, I think I have proven that there must be something more.

I think, moreover, that we may assume it to be proven that whatever this 'something more' is to consist in, it must be invariant as to the distinction between plain and significant narratives, and cannot accordingly be used to distinguish *between* plain narratives and significant ones. Our problem, then, is to find out what this something more is, and then, when we find it, to find what *further* thing will serve to sort narratives into the two classes. And here, concerning just this latter part of our problem, I should like to point out two things which the difference cannot consist in, if we are to suppose that the distinction between plain and significant narratives is to be a distinction *within* history, so that some historians will be writing plain, and others significant narratives; or that some will be writing narratives more plain (or more significant) than others, and still be doing *history*.

First, a significant narrative must be something less than a substantive philosophy of history, for a contrast exists between history and philosophy of history, and if a significant narrative were an instance of the latter, the intended contrast would not be a contrast *within* history. This is so even if a speculative philosophy of history, such as Hegel's, contains an ordinary historical narrative as part of itself (as Hegel's does). There can be little doubt that some statements which occur in philosophies of history might (indeed must) also occur in ordinary pieces of historical writing, since philosophies of history are concerned with the *whole* of history, including the past. Notice that philosophies of history *do* try to give explanations of the events they describe, and to attach some meaning to these events as well. So presumably the kind of explanation, and the kind of meaning relevant to genuinely significant narratives (which remain within history) must be different from these. I do not mean incidentally to argue that historians may not, wearing, so to speak, another hat, engage in philosophical speculative history. I only mean to say that when they do this, they are doing something *outside* history. If, finally, significant narratives were merely identified as speculative philosophical narratives of the whole of history, the distinction between plain and significant narratives would be between history and something else: not a distinction within history.

Secondly, there may be some theoretical works in the social sciences which contains, as parts of themselves, straight historical narratives. A book on business cycles may pause and tell a story. Nevertheless, a significant narrative cannot be this sort of story *together* with the rest of the book, because the book taken *in toto* is not a narrative of any kind, even if it happens to have a narrative part. If we were to call these books significant narratives, we would be doing more than abusing the notion of narrative, however. We would be contrasting plain narratives with works of a wholly different genre, contrasting history with the social sciences, and this, not being a distinction within history, would be inappropriate.

Bearing these limitations in mind, we turn now to the main question of seeking to determine by what criterion one might plausibly effect a distinction between kinds of narratives within history. It is true that Walsh has said that one kind of narrative explains, where the other kind merely describes. But he has gone on to say a good deal more than this, and what he says has a considerable interest in itself.[1] I want to examine his views, even though I shall ultimately reject his distinction, for we can, I think, learn a great amount about history in this way. *Enough* about history, in fact, to enable us to reject the distinction between history and chronicle, or between plain and significant narratives, or, what comes to the same thing, between explaining and describing in historical narration.

Walsh proposes that the difference between plain and significant narratives corresponds to, or represents, (1) two different levels of understanding and (2) two different kinds of knowledge. I shall consider these separately.

(1) Walsh offers, as instances of chronicle and history respectively, the sort of account we can, in the light of the information available to us, give of Greek painting in contrast with nineteenth-century political happenings. Thus illustrated, the 'distinction does answer to a real difference of historical understanding'.[1] The difference, in fact, is 'so profound that they might almost be said to constitute different genres'. Thus

> The narrative we can construct of nineteenth-century political history is both full and coherent; events in it can be presented in such a manner that their development seems to be orderly and intelligible. . . . But a history of Greek painting, or what passes for such a history, is a sorry affair by comparison, consisting of little but the names and approximate dates of a few celebrities, with the titles of their works as recorded by ancient authors . . . really an unsatisfactory chronicle, a mere skeleton of a history.[2]

Now if this is a correct description of the level of understanding we have of Greek painting, it seems obvious to me that the distinction wanted cannot be supported by these illustrations. For should an account of Greek painting be but 'a bare recital of unconnected facts', then, obviously, we have no narrative account at all of Greek painting. A list is not a narrative: the New York Telephone Directory is not a piece of historical writing, though it might have its uses were someone to set out to write the history of New York City. Consider a similar example. Contrast a table of the important and the secondary painters of the Italian Renaissance with a full history of Italian Renaissance painting. Here we would be contrasting, not two narratives, but a table and a narrative. As we would be if we were to compare a table of the Kings of England with a history of English Royalty. But now suppose that all we had, by way of information on Italian Renaissance painting, was a list of names and dates of painters and pictures. This would correspond to our alleged portion of information regarding Greek painting. Were this our situation (and it happens not even to be our situation with Greek painting), it would hardly follow that we could not write a narrative of Italian Renaissance painting. It is only that we could not adequately support, at every point, whatever narrative we might produce. And what is being overlooked in Walsh's analysis is the creative activity of what I shall term the 'historical imagination'.

References to imagination, in philosophical discussions, are almost certain to sound pious and windy. But there is, here at least, a logical point which this reference may bring out. To begin with, nobody need construct *de novo* the narrative history of the nineteenth century. It was an age permeated with an historical self-consciousness; men recorded, in narrative form, the events they were living through; and some of the greatest of its statesmen were amongst the greatest of its historians as well. We have inherited this, and our task has been to expand and modify, to correct and extend, this inherited account. We stand, perhaps, to this body of written history in something like the relation in which Lagrange stood to Newton. We are not so much obliged to invent a whole new theory as we are to tidy up and render elegant what is already received as a theory.

With Greek painting (or with our imaginary example of Italian painting), this is not the case. The Greeks did not see fit to write their own histories of art (which in itself tells us something about Greek painting), though bits and pieces of information about their art crept into the histories they did write, as well as into other writings. So here we are obliged to carry through a bit of imaginative reconstruction, to invent a theory, so to speak, in contrast with having only to polish up a theory already given to us. Too often, it seems to me, philosophers who have studied science have regarded science as a finished business, as a body of propositions already available which can now be reconstructed or rationally translated into some philosophically favoured language. And this tends, often, to induce a philosophical neglect of what has been called the logic of scientific *discovery*. But a comparable point may be made with regard to history.

Let us protract for a moment this analogy between a theory and a piece of historical narration—an analogy we have already considered in the discussion with Beard. We are, I think, entitled to suppose that a theory is logically distinct from the evidence that supports it. But then we might also say that a narrative is logically distinct from the evidence that supports it: footnotes are not proper parts of a story, but rather support the story at various points with evidence. It is true that historians might hesitate to publish a narrative which they were unable to support at every point. The historian might say, at a certain point, that he is resorting to conjecture: but this would mean a break in the foot-notes rather than a break in the narrative. At all events, a narrative is not just a summary of its own scholarly apparatus. It operates, instead, as a proposed account of what happened, and it can hardly be denied that such an account, operating as an hypothesis, might go to suggest support in favour of itself which was not initially available. There is *that* much truth in the view, considered earlier, that a statement about the past is a covert prediction of the outcome of an historical inquiry. But the relationship between a narrative and the materials which initially support it is, in a sense familiar to students of Peirce, *abductive*.[1] And in an import-ant sense, we cannot really make historical sense of whatever bits and pieces we may possess of 'history-as-record' until we are able to find a narrative for them to support. Indeed, until we have a narrative for them to support, it is something of a misnomer to regard them as evidence.

There are many sources from which support for a story, as well as suggestions for a story, might be drawn. In addition to actual records and documents, we almost certainly rely upon what we might call *conceptual* evidence. Simply to identify someone as an artist, for instance, already locates that individual under a concept, and permits us, with some measure of plausibility, to apply a whole set of different and, in the sense of the last chapter, *acceptable* or *possible*, sentences to that individual. What I wish to suggest is that these concepts not merely function as criteria of plausibility for narratives already written, but provide, as well, some basis for constructing a new narrative; in this particular case, a narrative of the life of someone identified as an artist. This narrative will be plausible to the extent that it tells us what might typically happen to an artist in his lifetime. Imagine that we had only this information: an artist, Leonardo da Vinci, lived in Florence at a certain date, and painted *The Last Supper*, a fresco in Milan. That the names of artists should have been recorded at all indicates a certain attitude towards painting: societies seldom leave lists of their cobblers or chimney-sweeps. That da Vinci should have been mentioned in such a list indicates that *he* was worthy of mention, for not *every* artist gets mentioned in such lists. The fact that the *Last Supper* is mentioned suggests that it was regarded as of some special importance, since it is the only painting by him mentioned (it is not plausible to suppose he did just one painting), and he is identified as the painter of it. To be sure, there is a problem as to whether the question asked was 'What did Leonardo paint?' or 'Who painted the *Last Supper*?' Whatever the case, we may suppose it be his most highly appreciated work and can assume it to have been his masterpiece. This, if we know the date of the picture, gives us some idea of his period of journeymanship, and whether he was a prodigy or not: knowing the life-dates of Masaccio, and the dates of the frescoes in the Brancacci Chapel *would* suggest he was a prodigy. Knowing the title of the painting and *understanding* it allows us to entertain some idea of the sorts of things it had to contain,[1] and we can suppose, as well, that if an artist of note painted a picture of note, and the latter had a religious motif, there was some more or less intimate connection between art and religion: at all events, we can get some general idea of who the patron was. These connections having been established, we can go on seeking for further ones, and for evidence to support the ones we had made. Bit by bit we would put together a *plausible narrative* of Leonardo's life. It would, to be sure, be a fairly general and schematic kind of account, and one might never have deduced from it, nor even have plausibly imagined from it, the sort of particular genius Leonardo was. It must not be overlooked that our actual knowledge of Leonardo's life has, in fact, entered into our concept of what an artist is, so that it is not easy to say what our concept would be if all we had to go on, in the case of Leonardo, were those few facts I mentioned. But the point is that we can stretch a few facts rather a long way, and that an imaginative appeal to our general concepts will fairly soon get us a narrative of some sort which we might then use as a guide for further research, seeing whether some further but independent evidence might be found for our narrative.

Without this further independent evidence (and here we *are* at the mercy of the resources of history-as-record), our narrative would float on air: it would, for all we knew, be fiction. But this surely helps us to see what is the difference between narratives and the evidence we have for them (a fictive narrative is one which requires *solely* conceptual evidence). One might say that the difference between a chronicle and a proper piece of history is the difference between a well-supported, in contrast with a poorly supported, narrative. And this, in turn, would suggest a comparison between a well-confirmed and a poorly confirmed theory. But this is not a difference between kinds or genres of theories, or, for that matter, between kinds or genres of narratives: it is a wholly quantitative difference between degrees of confirmation or amounts of support.

At one point Walsh suggests that, in addition to the decidedly limited amount of information currently available on Greek painting,[1]

There is also the fact that, because we stand nearer to the nineteenth century, we can enter into the thoughts and feelings of that age, and so use our evidence in a more effective way.[2]

This is an interesting view, in so far as it recognizes that it is not merely the amount of information we have, but also the way in which we use the information we do have that is important. There are, however, some highly controversial notions implicit in this statement. First, there is the implicit suggestion that we must make reference to the thought and feeling of human beings in order to make what they do appear intelligible. This is a rejection of behaviourism. Secondly, there is the implicit suggestion that we can succeed in such references, and hence succeed in making the behaviour of individuals intelligible inversely as the temporal distance between us and the individuals concerned increases. I shall not discuss the first of these. It is controversial to the extent that philosophical behaviourism is controversial, and, without for the present defending myself, I will register an agreement with Walsh's anti-behaviouristic stand for a limited number of cases. A defense would involve us in larger issues than those involved in understanding the past; issues, having to do with the understanding of *actions*. And if a philosopher were to say, as a behaviourist thesis, that we need never make reference to the thoughts and feelings of actors in order to understand their actions, it would follow from his stand that we would not need to make such reference to past actors and their actions, for his views are time-invariant. Walsh, of course, is by no means sceptical in general about either our ability or our need to make such references, but he does apparently subscribe to a qualified scepticism about past individuals, and this he does in proportion to the extent of their pastness. Therefore I shall address myself only to that suggestion.

There is one fatal objection against the thesis that increasing temporal distance entails decreasing intelligibility of human actions. It is that our temporal distance from the Greeks ought to make it equally difficult to write, or to understand, an account of *political* happenings in 300 B.C. and an account of artistic activities at that same period. But this simply is not the case. Thucydides' book is very nearly a paradigm of intelligible political history. His account, indeed, is so astute that we can apply it to our own time, and argue, if we wish, that people have changed very little. What it comes to, then, is *not* that we are temporally more remote from the Greeks of the third century B.C. than we are from, say, the French of the nineteenth century A.D.; but that we simply have a better understanding of political than of artistic behaviour. And this means that we have a more extensive and perhaps a more reliable stock of what I have termed *conceptual*, in contrast with *documentary* evidence for political constructions in contrast with artistic ones. One may then go on to suggest that our conceptual evidence, in the case of politics, will enable us to construct more complex narratives, independently of specific documentary evidence, than our corresponding conceptual evidence for artistic activity will permit us to achieve. If there should be any doubts about this, let someone imagine trying to write the history of nineteenth-century painting on the basis only of a list of names of artists and works. I dare say that relative proximity in time will help us very little. And if we knew all about the nineteenth century *except* what was painted, we could hardly *imagine* Impressionism.[1]

There are at least two difficulties with the notion of conceptual evidence which I had better remark upon.

The first is that it presupposes that the behaviour, towards the understanding of which it is applied, is invariant over time. And in so far as this is not the case, our use of conceptual evidence is decreasingly effective, not so much as a function of *time*, but as a function of the number and kinds of changes which may have taken place.[2] Just as the application of a scientific theory presupposes an isolated system, so, I think, the application of a narrative based upon conceptual evidence presupposes, in a somewhat similar way, constancy of institutions and practices. We could utilize conceptual evidence with impunity to reconstruct the history of a society, however long it might have endured, providing only that there were no changes in the sorts of practices covered by our concepts. But when this is not satisfied, there are some peculiarly historical difficulties. Ibn Khaldun stated them perfectly:

Now a dynasty will adopt many of the customs of its predecessors, while not forgetting its own, hence the prevalent set of customs will differ from the preceding generation. Should the ruling dynasty be supplanted by yet another, which in turn will mix its own customs with those prevailing, a new state of affairs will come about, which will differ from the first stage even more than it differs from the second.

This gradual change, in the direction of increasing difference, will proceed until it ends in total dissimilarity. . . .

Now men are naturally inclined to judge by comparison and analogy; yet these methods easily lead into error. Should they be accompanied by inattention and hastiness, they can lead the searcher far astray, far from the object of his inquiry. . . . Forgetting that great changes, nay the revolutions, in conditions and institutions which have taken place since those times, [they] draw analogies between the events of the past and those which take place around them, judging the past by what they know of the present. Yet the difference between the two periods may be great, leading to gross error.[1]

Now it may very well be that political behaviour has greater constancy over time than artistic behaviour does (think of the difference between the history of American politics in contrast with the history of American art over the past sixty years!), and this, then, would account for the difference in use of evidence which Walsh supposes. This, however, interests me less than do the special difficulties which apparently arise in connection with conceptual evidence. A narrative, for example, which does make considerable supporting use of it, and relatively little use of *documentary* evidence, necessarily depends upon certain general ideas which hold true, or are *held* to hold true, of the time at which it is written. Were our only evidence of *this* sort, all written history would indeed be 'present history'. This phenomenon we should want to call *temporal provincialism*. Certainly, it is a familiar enough phenomenon. One need think only of the great religious paintings, where the miraculous births, the adorations and annunciations, the passions and resurrections, are depicted as taking place in the Umbrian landscape, with Italianate donors looking on. A narrative which depends much upon conceptual evidence, has an inevitable contemporary or timeless ring about it, as though it were not about the past but the present, or not about a given time, but any time.

We are, I think, all of us temporally provincial with regard to the *future*. And this, in part, because for the events of the future, we have *only* conceptual evidence, and no documentary evidence at all. This will be an important fact for us later on when we shall consider the question whether we can write the history of events before they have happened. We can, of course. But we would hardly be able to support them, as we can narratives about the past, with documentary evidence, and, for this reason, our conception of the future has a curiously open and curiously abstract quality about it. If it is possible to make mistakes of the sort Ibn Khaldun mentioned regarding statements about the past, it is *a fortiori* possible to make mistakes about the future, for the precise reason that we lack the controls, which Ibn Khaldun was doubtless referring to, over the narratives we project—the controls of documentary evidence now available. This is, I should think, the final importance of history-as-record. Without it we would live quite wholly in the present, and have no idea that the past would or could have been different. But this too connects with our temporal provincialism regarding the future. For our conceptual evidence has to be modified in the light of documentary evidence, or, rather, narratives based upon the former need to be modified in the light of that latter sort of evidence when it can be found. But this helps show, and indeed gives some inductive grounds for saying, that conceptual evidence will not carry us very far. For if we are obliged to adjust it in the face of documentary evidence, so that we can say that concepts themselves have changed, surely we might expect as much for the future ? Thus future concepts will be as dissimilar to ours, as ours are to past concepts. We can *expect* that the future will be different from what our conceptual evidence would lead us to expect. That it *will* be different we can suppose. But in what way different is hard to say for not only do we lack documentary evidence : our conceptual evidence itself is not even *generally* adequate. If it is not for the past, why should it be so for the future ? These, then, are the limits of conceptual evidence, and if it were all we had, our conception of the future would resemble our conception of the past, and each would resemble our concept of the present. But this is to say that we would have no *historical* sense of the past or the future, and we would think timelessly. Narratives, then, based solely upon conceptual evidence would truly be historical and schematic, in contrast with our actual narratives of, say, political happenings in the nineteenth century. So in the end a narrative concerning the history of painting in Greece, eked out of the scraps of available documentary evidence plus whatever conceptual evidence we may have, would, after all, make a sorry contrast.

A second, and related difficulty, is this. Suppose we had a list of artists, together with their dates, and the titles of their works, but all of their works bore the title 'The Last Supper'. Using only conceptual evidence as a basis for constructing narratives of their individual lives, these narratives, in so far as they were supported by conceptual evidence, would be remarkably uniform. No statement could be included in one which could not, with equal justification, be included in another, and in all the rest. They would differ only in regard to name and date. This, of course, is just what we would expect, since all we could manage justifiably to say about any individual would have reference to just what he might have in common with every other individual covered by the concept. So unless, and until, we had further documentary evidence, we would have no way of justifiably individualizing the monotonously similar narratives. To be sure, we could arbitrarily *make* some differences, but we could not *justify* making them in one case rather than another. Meanwhile, the narratives based upon merely conceptual evidence would have the special abstract quality remarked upon before: they could be true of *any* artist (or any artist of that period) in the way in which '*x* was born, and sometime later, *x* died' is true of any man no longer alive. Now I think it is clear that historians are not concerned to write such abstract narratives. They are, rather, interested in writing indivi-duated narratives, narratives, if you wish, which are true of at most one individual. It is always, of course, a problem whether a given narrative (or a given description) we happen to produce is, in fact, true of at most one individual.[1] But this need not concern us here. What need concern us, rather, is that the distinction between an 'abstract' narrative and an individuated one does not represent the intended difference between plain and significant narratives. Every narrative, produced by any historian, intends to be an individuated narrative. In this sense, I think, significant narratives, no less than plain ones, would have it as their aim to say what really happened at a certain place and time, and this would not be altered by any further distinction we would want to make between kinds of narratives. Notice, however, that a narrative of Greek painting, based mainly on conceptual evidence, *fails* to do this. Thus if that sort of narrative is taken to be a chronicle, or a plain narrative, we could hardly then characterize chronicles, or plain narratives, as being 'an exact description of what happened'.

We might, then, say that we have two different levels of understanding here. But this does not correspond at all to a difference between saying what exactly happened, and then doing something more than that. It corresponds, rather, to the degree of individuation, which is the consequence of the different amounts of documentary evidence, we are able justifiably to give to our narratives. A history of Greek painting is clearly less individuated than a history of political happenings in the nineteenth century.

(2) At one point Walsh suggests that chronicles have the same relation to history as sense perception has to science. There are, of course, many different kinds of relations between sense-perception and science, but I suppose the most natural interpretation of Walsh's suggestion is this: the difference is comparable to the difference between perceiving that something is the case, and explaining why it is so. Certainly one could not accept the suggestion that the difference is instead comparable to the frequently discussed contrast between the common-sense, and the so-called scientific description of the world. For this seems quite inapplicable to history, and certainly inapplicable to the specific examples Walsh himself provides us with. The history, for instance, of nineteenth-century political happenings would just be an instance falling within the ordinary or common-sense view of the world: it describes people and their actions in just the way in which we would ordinarily describe them, and this, in part, because such narratives are written in the ordinary language we all speak, and with which we express the so-called common-sense view of things. If anything, a narrative concerned with Greek painting would be a shade *more* remote from

common-sense views, but this would be due to the fact that common-sense (as we saw in our discussion of conceptual evidence) is better able to deal with political than with artistic behaviour. Nevertheless there is little difference, for the language of historical narration is seldom technical, in the way in which scientific vocabulary is, and most literate persons would be able, without having to acquire any special vocabulary or any special skills, to follow narratives of nineteenth-century political behaviour. Indeed, they would very likely have to master a great deal *more* special language in order to follow a narrative of Greek painting. So let us consider simply the natural interpretation of Walsh's suggestion.

There is, I think, little doubt that one can draw a distinction between perceiving that x is the case, and explaining why it is so. Here, to be sure, one would want to make some careful distinctions. One man might say that he sees a blinding flash, while another, witnessing the same phenomenon, would say that he saw a magnesium explosion, and the latter description of the same phenomena is very nearly an explanation of what was seen. Nevertheless, complexities of description to one side, one might surely agree that there is a difference between saying only that Napoleon lost at Waterloo, and going on to explain why he did lose. The only difficulty here is that we are concerned to find a difference between two kinds of *narrative* and 'Napoleon lost' is not a narrative. But now it might be argued that we could nevertheless have one narrative which merely described what happened, and another which explained why it did happen. I wish to argue, however, that a narrative which *fails* to explain is very likely to be a statement like S, and hence not really a narrative; while a narrative which does explain does exactly this: it says what actually happened, and so qualifies as a *plain* narrative by Walsh's criteria. We would be left then with the problem of finding out what a significant narrative might do which is different from this. And I shall say there is nothing it can do, beyond saying what precisely happened, so long as it is to remain an *historical* narrative. The distinction, then, is not one which may be made *within* history.

Plain narratives, writes Walsh, have as their purpose to report 'in the famous phrase of Ranke's "precisely what happened" and leave the matter at that'.[1] There is apparently some difficulty in interpreting

Ranke's claim that his history wants to show what actually happened (*wie es eigentlich gewesen*).[1] He himself was only pointing a contrast: he did not aspire, he says, either to judge the past or to 'instruct the present for the benefit of future ages'. He was only interested in saying what had actually happened. Even so, men have found this originally humble disclaimer an extraordinarily boastful claim, and one which a man cannot live up to. He has been understood to mean, for example, that nothing of himself should be revealed in his wholly objective history;[2] or that everything about his subject should be mentioned in it.[3] And both these things, it has been suggested, are impossible. Let us just consider the latter rendering. It is, I dare say, true that one could not at once obey the command to give an account of some happening, and the command to mention *everything*. Accounts, I have argued, must by their nature leave things out, and in history as elsewhere it is the mark of someone capable of organizing a subject that he knows what to exclude, and is able to assert that some things are more important than others. Suppose I wish to know what happened at a court trial. I may ask my informant to leave nothing out, to tell me all. But I should be dismayed if, in addition to telling me of the speeches of the attorneys, the emotional attitudes of the litigants, the behaviour of the judge, he were to tell me how many flies there were in the courtroom, and show me a complicated map of the precise orbits in which they flew, a vast tangle of epicycles. Or mention all the coughs and sneezes. The story would get submerged in all these details. I *can* imagine him saying: 'At this point a fly lighted on the rail of the witness-box.' For I would expect something odd and interesting to follow: the witness screams, displaying a weird phobia. Or a brilliant attorney takes this as an occasion for a splendid forensic display ('As this fly, ladies and gentlemen . . .'). Or in trying to brush him away, a bottle of ink gets spilled over a critical bit of evidence. Whatever the case, I shall want to know: what *about* that fly? But if there is no 'what about', if this is only 'part of what happened during the trial', then it does not belong in the account of the trial at all. When I say, then: 'tell me the whole story, and leave nothing out' I must be (and am) understood to mean: leave out nothing significant: whatever belongs in the story I want to be told of it. And this, surely, is what Ranke must have meant in the main.

There are few problems in philosophy which merit closer analysis than the question of relevance, but here I shall be appealing only to our intuitive ideas, in accordance with which we are able to recognize something as belonging or not belonging to a certain story: even a child can do that. If telling what precisely happened means what some critics of Ranke seem to think it means, what Ranke would ideally have produced would not even have been a *plain* narrative: for it would not have been a narrative. I shall say, then, that any narrative is a structure imposed upon events, grouping some of them together with others, and ruling some out as lacking relevance. So it could not be a distinguishing mark of any given kind of narrative that it does this. If you wish to put the matter trivially, you can say that a narrative mentions only the significant events: but in this respect every narrative would be concerned with finding the significance of events: each narrative would ideally want to include only those things relevant to some other events, or significant to them. We could hardly divide narratives into classes by this criterion except, perhaps, into good and bad ones, where the bad ones contain some amount of insignificant detail.

It is not easy to see what kind of significance it could be that historians could attach to events which would go to make a philosophically important distinction between kinds of narratives. There are, for instance, various events or figures which are regarded as more significant than others. The Battle of Waterloo, let us say, was more significant than the Battle of Wagram, and Napoleon a more important general than Blücher was. In a parasitic sense, narratives of Waterloo and Napoleon might be more significant than narratives of Blücher and Wagram. But this is of scant philosophical importance, I should think, and at all events either narrative might tell precisely what happened. What might be philosophically important would be to specify some of the different senses in which we speak of an event or an individual as significant, and this I shall now proceed to do. I shall try to show, moreover, that some of these senses do involve historians in something more than saying precisely what happened. Therefore, I shall also try to show that none of these extra things really make the required distinction a philosophically important one.

(1) *Pragmatic Significance*. Sometimes an historian chooses a certain

happening or individual to write a narrative of because the subject has, for him, a moral interest, so that, in addition to writing what precisely happened, he hopes to be making a moral point of some sort. His narrative will then serve a certain purpose over and above telling us what really happened. Frequently, the historian's tone will show what moral point it is that he wishes to make. Gibbon, for instance, writes in a contemptuous tone of the excesses of Byzantine rulers. He meant, by so doing, to point a contrast between them and the rather more enlightened monarchs of his own age. There can be little doubt that some of the things he included in his book are specifically there because of his moralistic purposes. A converse point was made by Tacitus in *Germania*. Here he specifically chose to write of Germany in order to point an individious contrast with the behaviour, and in particular the sexual behaviour, of his own countrymen: he stresses thus the virtuousness of the Germans. One could multiply such examples. Histories of the lives of popes, or Captains of Industry, or courtly ladies of old Japan, will often have a significance in this sense, and such histories are specifically intended, and sometimes explicitly constructed, to serve a moralistic purpose. All such narratives might then be regarded as significant rather than plain. A relativist, of course, might wish to say that *all* narratives are significant in this sense, since all historians are dominated by some sort of moral purpose and pragmatic intent, and this serves to determine what sorts of things they write of, the way in which they write of them, and the events they regard as relevant. Whether this is so or not, the fact remains that we can at least conceive of narratives which do not, and Ranke, at least, claimed not to have such ulterior purpose: *he* was concerned to say only what really happened and, in this sense, to write a *plain* narrative.

(2) *Theoretical Significance.* A set of happenings may be significant to some inquirer because he sees them as standing in an evidential or illustrative relationship to some general theory he is concerned to establish or disestablish. Thus, the Cromwellian revolution may be taken either to confirm a general theory regarding revolutions, or to be a counter-instance of such a theory; and it is relative to some such theory that the event derives its significance. The particular narratives concerning French history by Marx are examples of this, serving to illustrate a general theory of class-struggle. A narrative of the same events, written

to rebut the Marxian theory, would be equally significant if we regard now 'significant narrative' as a narrative written for this sort of theoretical purpose. A plain narrative would then be one which had no such purpose. Once again, in a loose sense, every narrative might be significant from this point of view and even Ranke's narrative *might* be significant in the sense that it was explicitly written to show that objective history was possible: its significance would lie in its plainness.

(3) *Consequential Significance.* An event *E* may be said to be significant to some historian *H* when *E* has certain consequences to which *H* attaches some importance. This is, for instance, the psycho-analyst's sense of 'significance' when he finally says to a patient, in the latter's review of his past, that he has touched upon something significant. And it is far and away the typical use of the term in historical writings. When we say of an event that it has no significance, we mean, not that it has no consequences, but rather that it has no *important* ones. So this sense of significance is logically connected with an independent notion of *importance*, where the latter may depend upon any number of different criteria. Examples here are easy to find. We say that as a consequence of the Persian Wars, the Hellenic people, and particularly the Athenians, were able to develop along autonomous lines and to consolidate their cultural achievement. We say that the significance of the Black Death was that it created a sellers' market in labour, hence a rise in wages, hence contributed to the break-up of the feudal structure of tied labour. It is this sense of significance which is appropriate when, in accordance with a well-known *mot* of Pascal's, we say that the size of Cleopatra's nose was of historical significance. A narrative which describes or shows the significance of this or that event might be called a significant narrative. It is, on the other hand, difficult to conceive of a contrasting sort of plain narrative, for this notion of significance seems essential to the very structure of narratives. If an earlier event is not significant with regard to a later event in a story, it does not belong in that story. And one can always justify inclusion of one event by showing it to be significant in just this sense. If every pair of events mentioned in a story are so unrelated that the earlier one is not significant with regard to the later one, the result is in fact *not* a story, but rather a set of statements approaching *S*.

(4) *Revelatory Significance.* I have suggested that the relationship of a

story to a body of evidence may, at a certain stage, be abductive. That is, on the basis of some set of records, we *postulate* a kind of story, and then go on to seek out further supporting evidence. Such evidence, once found, may be regarded a significant find in that it finally supports a claim we were hitherto uncertain of. Now comparably, there may be some gap in a story, or a part of a story may be quite wrong, or there may be things which happened which we are unaware of and so do not feel that there is a gap in the story we have. And then we hit upon some record which tells of events which fill the gap, or are different from what we thought actually happened, or which tell us something we had not known. Such discoveries are significant because they reveal something heretofore unknown, and we might derivatively regard the events themselves as significant. This, of course, is relative to a state of knowledge: you cannot reveal things to people who are already aware of them, and yesterday's revelations are the stale news of today. Nevertheless, this is an important notion of significance, and I shall apply it in the following way to sets of events. I shall say that a set of events E is significant to an historian if, on the basis of them, he is able to reconstruct or somehow infer the occurence of some other set of events. Asked, for instance, what is the significance of Descartes' having moved to Holland, I might say that this event signifies the fact that there were forces afoot in France which were repressive of free thought, and that these forces were absent in Holland. *Postulating* this thesis, I might set about trying to verify the presence of such forces in France and the absence of them in Holland. Here again there is a psycho-analytical analogue. I might say that the significance of x having married an older woman is that he was seeking to replace his mother. One might now say that a significant narrative is one which narrates events, or sets of events, related in this manner. On the other hand, it is not easy to see what a plain narrative would consist of. For example, suppose E is significant of a set of events which stand in some *explanatory* role to E. I should think it odd if an event included in a narrative did not stand in an explanatory role to another event. For what after all is the relationship here if not this, that we help make sense of E by reference to some other event? And if none of the events mentioned in a narrative help make sense of any of the others, we again have something more like S than a narrative.

Now this list of senses of 'significance' is hardly exhaustive, and is perhaps not even exclusive: (3) might be merely a special case of (4); a narrative might be significant in both the sense of (2) and of (4); and so forth. But it will still serve my present purposes well enough, and I shall now proceed to comment briefly upon each of the entries I have made.

a. It cannot be denied that historians can and do find moral guidance, moral parallels, horrible examples, and moral paradigms in the events of the past. Nor can it be denied that their motive for writing history at all is frequently pragmatic: they wish to restore or ruin a reputation, to offer moral instruction, or to support or reinforce a moral position. Nevertheless, none of this is incompatible with reporting what exactly happened, and indeed, unless they do that, they are not writing history at all. To be sure, different historians, with differing moral persuasions and purposes, might write different stories. But then each might be writing precisely what happened despite this, for they are writing, in the end, about quite different things, and the only quarrel between them would *be* a moral quarrel, the subjects of their narratives being distinct. If, on the other hand, they both attempt to tell the same story, and the stories differ, their quarrel is not merely moral but factual. In that case, however, one or both of the narratives will be defective in the only relevant historical sense, namely, in failing to state precisely what happened. Assuming this is corrected, they may continue to disagree morally, but this is no longer an *historically* relevant disagreement, for they might disagree independantly of historical information, and, indeed, if they do agree on all the facts, their further disagreement regarding moral interpretations is plainly irrelevant to history, and history is irrelevant to it. What cannot be represented as a factual disagreement is irrelevant to history, and what can be so represented is nothing more than a disagreement over what precisely happened. A significant narrative, in the present sense, would be a plain narrative plus a moral interpretation. But it is the plain narrative which is history. The moral interpretation is extra-historical, so the contrast between a plain and significant narrative is not a contrast within history, but between history and something else.

Someone may, of course, argue that the so-called distinction between beliefs and attitudes is not clear. I should reply that to the same extent, the distinction between plain and significant narratives is not clear. Similarly, should someone argue that it is impossible to say what happened without, as a consequence of the very language we employ, making some moral judgement or other, then, of course, there just are and can be no plain narratives, so the distinction is non-existent. *Per contra*, should someone wish to argue that ethical predicates are *not* expressions of attitude, but describe real properties of things and events, and that any description which fails to use them is incomplete, then we might indeed allow a difference between plain and significant narratives. A significant narrative would succeed in what a plain one falls short of, namely to report precisely what happened. For a factual *plus* an ethical description would be a more accurate account, or a fuller one, than a merely factual account.

I myself think the distinction is clear enough. Were someone to say that here is a good piece of history though it makes no moral point, everyone would understand him. He would certainly be saying nothing inherently inconsistent. If he were to say it is a bad piece of history because it makes no moral point, he would not, in fact, be dispraising it by any normal criterion for classifying histories as better or worse. It is *not* generally accepted as a reason for saying that something is bad history that it is morally neutral, anymore than saying that it does not mention Napoleon is taken to be a reason for calling something a bad piece of history. On the other hand, if someone were to say that something is good history, though it does not say what happened, he seriously raises the question why anyone would call it history to begin with. To say that it is good history *because* it tells what happened is to offer the standard reason for calling a piece of history good, while to say, finally, that something is a good piece of history though it tells what happened is to say something verging on unintelligibility. But to say that it is good history though it makes a moral point is intelligible enough. It means that its making a moral point has not interfered with its satisfying the criteria of good history. We may, I think then, forget about sense *a*.

b. I have been endorsing the view that narratives may be regarded as kinds of theories, capable of support, and introducing, by grouping them together in certain ways, a kind of order and structure into events. A narrative, so considered, is nevertheless localized as to space and time, it forms an answer to an historical question, and is accordingly to be distinguished from a general theory which is not thus localized, and is not, therefore, an answer to an historical question. Sense *b* of 'significant' has application to narrative written specifically to illustrate or confirm some general theory which is not itself an answer to an historical question, but rather to a scientific, or to one kind of scientific question. To be sure, the identical narrative might have been written, whether the writer had this ulterior purpose in mind or not. We might judge it, as a narrative, quite independently of whether it serves or fails to serve any extra-scientific task it might be set to do. I would say, then, that illustrating or confirming a general theory is a non-historical task, and questions like 'Is it a good illustration?' or 'Does it confirm theory *T*?' are *not* answers to historical questions, though the narrative itself will certainly *fail* to do these things if it fails to be an answer to the historical question appropriate to it. Thus, if it fails to satisfy the minimal historical requirements, it can hardly do a higher- (or different-) order job. We do not, at any rate, have here two distinct kinds of narratives, but only one kind, though it is sometimes put to a non-historical use. The relationship between a narrative and a general theory may be profitably understood as similar to the relationship between a narrative and a moral thesis or purpose. I shall be wanting to make some further remarks on this relationship, but for the present, I think, we may regard sense *b* as irrelevant to any distinction within history, and as failing to provide us with a suitable distinction between plain narratives and significant ones.

c. Suppose the difference between a plain and a significant narrative consists in the fact that the latter spells out the consequence of some set of events, while the former simply relates those events. Let us call these two narratives *N-s* and *N-p* respectively. Now, *ex hypothesi N-p* is a narrative. Accordingly it must satisfy some further conditions than our paradigm non-narrative, *S*. But this means that at least some of the later events mentioned in *N-p* must be significant with regard to some of the earlier events, that is, those later events *are* the significance of the earlier ones in that they are their consequences. If, by mentioning no event mentioned by *N-p* we are able to answer the question, what is the significance of *this* event, asked of *any* event in *N-p*, then *N-p* fails to be a narrative. Accordingly, every narrative must spell out some consequences of some events, and the difference between *N-p* and *N-s* is one of degree. In view of this fact, it is hard to see, if *N-p* can be taken to report what precisely happened, why *N-s* should be differently characterized. It may report *more* of what precisely happened than *N-p* does, but this is not the same thing as doing more than reporting precisely what happened. On the other hand, if *N-p* fails to report precisely what happened *because* it spells out some consequences of some events, then it is hard to see how a narrative *can* report precisely what happened. The intended contrast would not, accordingly, be between kinds of *narratives*. By sense *c* of 'significant,' then, every narrative is significant. But then, if one narrative should happen, in whatever sense is acceptable, to describe precisely what happened, and the other does not, then the latter is, in so far as it does not describe precisely what happened, disqualified as history. From this it follows that every narrative in history is a *plain* narrative. So every historical narrative is indifferently plain or indifferently significant.

d. We may deal quite briefly with *d.* Suppose we have a gap in a narrative *N-p*, and for want of available documentary evidence, we are unable to fill it except by resorting to some sort of conceptual evidence. We know, say, that *E-1* and *E-3* happened, and we have the sense that they are connected, but we do not know what the connection is. Notice that the gap here is relative to a narrative organization. Let us postulate an event *E-2*. Now, however, a revelatory piece of evidence is discovered, on the basis of which we can fill this gap in and close the narrative, so to speak. Our new narrative happens to succeed where *N-p* failed, namely in reporting what exactly happened. This in general is what discovering something of revelatory significance helps us to do. In revealing something we had not known before, or only suspected, it enables us to report what happened more precisely than we would have been able to do without it. Revelatory significance will not thus effect the required distinction.

For these reasons, it seems to me fair to say that there are not two kinds of narratives in history, or at least not two kinds of the sort we have been discussing here. Ranke's characterization, whatever its vagueness, and whatever implausible readings of it may have been given by unsympathetic critics, is an admirable characterization of what historians seek to do. Indeed, I might regard it as a variant statement of what I have called the minimal historical aim. Yet, in the sense in which historians describe what happened by means of narratives, they are, since a narrative itself is a way of organizing things, and so 'goes beyond' what is given, involved in something one might call 'giving an interpretation'. Presumably there are problems which arise in connection with the semantical connection between narratives and 'history-as-actuality', and the truth-conditions for narratives are apt to be complex. But so far as genre is concerned, history, I am saying, is of a piece. Any kind of narrative, assuming there were kinds of narratives, would require and presuppose criteria of relevance in accordance with which things would be included and excluded. This means, I think, that the maximally detailed account, that ideal duplicate of history-as-actuality, would not be a narrative.

Professor Walsh has argued,[1] in one place, that there is a difference between establishing a fact and establishing a connection between facts; that these two kinds of activity are on distinctly different levels. One might, I suppose, say that the notion of a fact is not clear, but that it is a *fact* after all, that two things are connected. Yet it might be argued that there are levels of fact. To establish that *Ei* happened, and then to establish that *Ej* happened, is to do something philosophically distinguishable from establishing some connection between *Ei* and *Ej*. On this point I wish to make a few comments.

(1) It is certainly true that there is, in historical practice, such a thing as establishing it as a fact that a certain event occurred. The extent to which this can be done without having established connections between this event and other events in the past is not easy to say, and I am inclined to believe it could not be done at all. This, however, I let pass, because someone might, I think, merely be interested in establishing that a certain picture, say, was painted at a certain date, and not be interested in telling any story. Suppose the historian establishes that the painting was done in 1817, and publishes a paper showing this. The paper might not be a narrative at all, though doubtless a narrative will be presupposed, and though the newly established fact may eventually enter some narrative. Nevertheless, if the paper is not a narrative, it is not a plain narrative. The historian has indeed answered an historical question. He has made a true statement about the past. What he has done, however, is not to be understood as establishing a narrative; and to contrast his work with historians who do write narratives is *not* to make a contrast between kinds of

narratives. I have been concerned only with whether there are kinds of narratives to be contrasted.

(2) How is one to write a narrative without establishing connections between events? To contrast an account which connects events with an account which does not, is hardly to be contrasting a narrative with a narrative, but rather a narrative with something quite different, something like S.

(3) We sometimes have a Humean tendency to think of events as discrete and pellet-like, and to think of connections between events as not consisting of some intervening pellet. I shall not quarrel with this view here. What I do wish to insist upon is that not every true description of an event can be made by means of monadic predicates solely. Similarly with descriptions of things. It is a true description of my typewriter that it is black; it is also a true description of it that it is on the table in my room, and it is also true that it is the machine on which I wrote a letter five days ago. To establish some descriptions of things or events *requires* that we establish connections between them and other things and events. Here, for instance, are a pair of descriptions of the same event.

D-1. Jones struck a match.
D-2. Jones revealed the position of his squad to the enemy, inadvertently destroying the tactical advantage they had enjoyed.

One cannot establish D-2 without establishing a whole host of connections with other events, and some of these connections cut across time.

(4) It might be agreed that all narratives connect events. But then, it might be pressed, some do more than this. They *explain* in addition to reporting precisely what happened. It is this that makes the difference between plain narratives and significant ones. The trouble with this suggestion is that it overlooks the extent to which a narrative is already a *form* of explanation. To contrast narratives with other forms of explanation may be important, but this is not the required sort of contrast. A narrative describes and explains at once.

(5) There are descriptions of the past other than narrative ones. This does not, of course, help the distinction, but it raises some interesting questions. I shall be concerned with this question only. Narratives, by definition, leave things out. Yet if one did not use the narrative form, one

might be able to give the complete description ideally presupposed as the aim of history, and so effect the statement ideally at the nether end of the range of statements marked out by an historical question. I shall say now, that you cannot give a complete description of any event which does not use narratives. Completely to describe an event is to locate it in all the right stories, and this we cannot do. We cannot because we are temporally provincial with regard to the future. We cannot for the same reasons that we cannot achieve a speculative philosophy of history. The complete description then presupposes a narrative organization, and narrative organization is something that *we* do. Not merely that, but the imposition of a narrative organization logically involves us with an inexpungable subjective factor. There is an element of sheer arbitrariness in it. We organize events relative to some events which we find significant in a sense not touched upon here. It is a sense of significance common, however, to all narratives, and is determined by the topical interests of this human being or that. The relativists are accordingly right. All of this I shall try to show next.

VIII

NARRATIVE SENTENCES

I mean to isolate and to analyse here a class of sentences which seem to me to occur most typically in historical writings, although they appear in narratives of all sorts and may even enter into common speech in a natural kind of way. I shall designate them as 'narrative sentences'. Their most general characteristic is that they refer to at least two time-separated events though they only *describe* (are only *about*) the earliest event to which they refer. Commonly they take the past tense, and indeed it would be odd—for reasons I shall want to consider here—for them to take any other tense. The fact that these sentences may constitute in some measure a differentiating stylistic feature of narrative writing is of less interest to me than the fact that use of them suggests a differentiating feature of historical knowledge. But even this is less interesting to me than the fact that narrative sentences offer an occasion for discussing, in a systematic way, a great many of the philosophical problems which history raises and which it is the task of the *philosophy* of history to try to solve. Indeed I shall introduce them in the context of some of these problems. My thesis is that narrative sentences are so peculiarly related to our concept of history that analysis of them must indicate what some of the main features of that concept are. In addition they help show why the proper answer to the tedious question 'Is history art or science?' is: 'Neither'.

Peirce wrote to Lady Welby: 'Our idea of the past is precisely the idea of that which is absolutely determinate, fixed, *fait accompli*, and dead, as against the future which is living, plastic, and determinable.'[1] Certainly this *is* what most of us think. But could we hold a different view? For a variety of reasons, some men have held that the *future* is fixed and determinate as well as the past. Suppose all we know about Caesar is that he existed. Whether or not he was even in any particular place, say England, is not known. Yet we might appeal to a venerable notion, the Principle of Excluded Middle, and say either that he was there or that he was not there, and that at least one of these alternatives is true. Why might not someone in the fifth century B.C. have invoked the identical Principle to argue that Caesar either will be in England or not? Perhaps because nobody then could have known that Caesar would exist the way we know that he *has* existed. Still, he might have said that either Caesar would exist or not, and that one of these statements would have to be true. *If* the Principle may be invoked for this future matter of fact, why not for all? What, however, could the name 'Caesar' mean to such a person, just what sort of thing is it he is saying will exist or not? Well, I have supposed that all *we* know is that he existed. Doubtless this is unrealistic. But flesh a description out as we will, what is to have prevented a fifth-century B.C. speaker from saying that someone of just that description would exist or not? Should he have spoken in this way why should the Principle not guarantee that at least this description or its negate is true? Or does it hold only for the past? After all, there are *four* possibilities, including the possibility that the future is determinate and the past 'living, plastic, and determinable'. Why is it that our 'idea' of past and future corresponds only to the possibility which Peirce described? Granted that this is our idea, the question remains *why*.

Our natural temptation these days is to say that it is a matter of definition. Consider, however, the wild fantasy of the whole course of history going suddenly into reverse, like a film strip running backwards. After a time would come the sound 'thgil eb ereht teL' and darkness would once again settle upon the face of the waters. The future would then be the exact mirror-image of the past, and there would be a rule by means of which an exactly corresponding sentence about the future could be found for every true sentence about the past. In such a case the future would be on an exact footing with the past in point of determinateness. True, we cannot put ourselves in this picture: nobody could *know* that what was happening was the reversal of history—for this would destroy the symmetry. Perhaps what we mean by the indeterminateness of the future is that we *can* put ourselves in the picture, there is room for us to move. But for that matter we can, in imagination at least (and this is all that matters here) put ourselves into the past, as in *A Connecticut Yankee in King Arthur's Court*. In fact, of course, there is no room in Arthur's England for twentieth-century strangers. However, there would also be no room in the corresponding segment of the future were history to reverse itself. No-one is saying that history will do this: but it is not a matter of *definition* that it will not.

Let us say that we are empirically certain that the future will not be the image of the past. What then *will* the future be like? People may guess this and that, but in contrast with our knowledge of what has taken place, we are very uncertain indeed with regard to what will happen. Can this be what is meant by the past being determinate and the future merely determinable? So that our 'idea' is based not upon some definition of past and future but rather upon our knowledge of each? Then Peirce's statement is false. We are always revising our beliefs about the past, and to suppose them 'fixed' would be unfaithful to the spirit of historical inquiry. In principle, any belief about the past is liable to revision, just in the same way perhaps as any belief about the future. Actually we are sometimes more certain about the future than we are about the past. At a given moment I am far more certain where a falling pine-cone will land than I am with regard to where it fell from. At best the difference is one of degree.

Peirce also wrote 'the existent is determinate in every respect'.[1] Possibly what we want then is a kind of ontological interpretation of his original claim. The future, if it is not determinate, does not exist. But if the contrast is to work, the past must exist, however this is to be understood. This may even take care of the Principle of Excluded Middle! Since there is nothing for sentences purportedly about the future to refer to, the question of their truth or falsity fails to arise.[2] Or we might say: the past has been constructed, but the future has not, and so make a somewhat punning extension of Intuitionism to get rid of that vexing Principle.[3] Of course it would hardly do to say that our idea of the past is of something existent and our idea of the future of something non-existent. If anything, our idea of the past is of something that *has* existed while our idea of the future is of something which *will* exist. Very few people think that the past exists. But some very good philosophers have thought this way. 'It appears to me that, once an event has happened, it exists eternally,' writes C. D. Broad.[4] For surely, he argues, we can meaningfully say that a certain event is past, that is, stands in a certain temporal relation with some other event. But if it did not exist, the relation would collapse for want of a term, and our statement about it would be nonsense. So all such events must constitute a 'permanent part of the universe'.[1] This seems a very weak argument indeed to support so vast a consequence, and we might as consistently argue, *mutatis mutandis*, that if we can meaningfully say that a certain event is future, that event must either exist eternally or all statements about the future are nonsense. But let us suppose that Broad is right, and let us manufacture a metaphysical model to satisfy our idea about the past and future which this interpretation of Peirce's statement of it seems to require. The important feature of this model is the fixedness of the past. Notice that this metaphysical excursus does not explain *why* we have the idea that the past is fixed and the future fluid. It only shows what the world must be like if our idea is to be *true*.

Let the Past be considered a great sort of container, a bin in which are located, in the order of their occurrence, all the events which have ever happened. It is a container which grows moment by moment longer in the forward direction, and moment by moment fuller as layer upon layer of events enter its fluid, accommodating maw. The forward lengthening of the Past is irrepressible, and regular; and once within the container, a given event E and the growing edge of the Past recede away from one another at a rate which is just the rate at which Time flows. E gets buried deeper and deeper in the Past as layer after layer of other events pile up. But this constantly increasing recession away from the Present is the *only* change E is ever to suffer: apart from this it is utterly impervious to modification. E, moreover, will generally be but one of a set of events which enter the Past together. In this case, E and its contemporaries constitute an exclusive class, in the sense that no *further* event will ever join them as, so to speak, a new contemporary. So the Past is not to change either through any modification of E apart from its momently increasing pastness, or through the addition of some other event contemporary with E which E lacked as a contemporary upon its entry into pasthood.

This 'model' construes events as time-extended entities in a Universe extended in time, a view conceivably licit. What is *not* quite licit in the model is that part of it which suggests that E and its contemporaries are exact co-evals, having each the same amount of temporal thickness and coincident termini. Common use of the term 'event' is fairly chaotic, and we are likely to apply it to occurrences of varying duration, even null duration. Seeing a robin, for instance, is perhaps an important event in the bird-watcher's morning. But such an event might be classed with what Ryle has called 'achievements', and can, in his phrase, be dated but not clocked.[1] We can both date and clock events such as flashes of lightning. We speak of the French Revolution or the Civil War as major events in the history of France and America respectively, and these are better measured by the calendar than the clock, provided we agree where in time to begin. Fidelity to usage requires us then to think of events as of varying duration, the only alternative being arbitrarily to decide that an event is exactly so long, say three minutes.[2] But if we follow usage, we may be obliged to say that E, though it may have many contemporaries, might still have no precise co-evals, so that a line drawn perpendicular to the direction of time at the anterior terminus of E would conceivably not intersect the anterior terminus of any of E's contemporaries. This, however, has untoward consequences for that part of our model which has events piling up, layer after layer, and proceeding away from the present in an orderly manner. For suppose that E has wholly entered the Past while its contemporary E' has only partially achieved pasthood, having part of its career yet to run. One may now ask *where* is the rest of E' when that part of it which overlaps E is in the Past. Somehow one feels uneasy thinking of it protruding like a worm half buried in a can of dirt. True, we can say that part of it which is not in the Past is in the Future, E' merely passing from one container into another. But then suppose that E' overlaps both E and E'' though neither of these overlaps the other. Then when E is wholly in the Past, E'' is wholly in the Future. But then the Future exists after all, and the desired contrast between the determinateness of the Past and the indeterminateness of the Future falls through. No, we shall have to say that the rest of E' does not exist. But suppose the 'rest' of E' fails to happen? Well, then, the Past must contain fragments of events as well as events. With this lame addendum, we may continue to employ the model, for what it is worth.

I admit that it is not worth much. The Future, for one thing, has been dealt with very casually. But anyway, 'there', in the Past, are situated all the events which ever have happened, like frozen tableaux. They are stowed in the order of their happening, they overlap (for they are of varying sizes) and interpenetrate (for an event E may have another event E' as a part of itself). More importantly, they cannot change, nor can the order amongst them change, nor can the Past acquire fresh contents save at its forward end. *Why* they cannot change is not yet clear. But there must be strong reasons, for according to an old tradition not even *God* can undo what once has been done: '*Niente diminisce la sua omnipotenza il dire che Iddio non puo fare che il fatto non sia fatto.*'[1] But I shall leave that problem for the time, and turn to the matter of describing our inert Past.

By a *full description* of an event E I shall mean a set of sentences which, taken together, state absolutely everything that happened in E. Since the sequence of happening is important, we should want this order reflected in the full description by some device or other. Indeed, a full description will be an order-preserving account of everything that happened. As such, a full description bears some analogy to a map: there is an isomorphism between the full description and the event of which it is true. Now with maps there are two sorts of problems. First, there are things in the mapped territory that are not designated in the map, so that in common practice maps are incomplete, and do not exactly duplicate the territory.[2] Second, maps go out of date because territories change: coast-lines get washed away, cities are destroyed and others spring up, boundaries are drawn afresh as a consequence of wars and treaties.[3] This second problem does not arise for full descriptions of past events inasmuch as the Past does not change. But then neither need the first problem. We can imagine a description which really is a full description, which tells everything and is perfectly isomorphic with an event. Such a description then will be *definitive*: it shows the event *wie es eigentlich gewesen ist*. The maps of all the events may now be supposed assembled, to constitute a (really *the*) map of the whole Past. This global map then changes only in the way the Past itself changes: it is added to along the forward edge. It now hardly matters whether we talk about the Past or its full description.

Narrative Sentences

I now want to insert an Ideal Chronicler into my picture. He knows whatever happens the moment it happens, even in other minds. He is also to have the gift of instantaneous transcription: everything that happens across the whole forward rim of the Past is set down by him, as it happens, the *way* it happens. The resultant running account I shall term the Ideal Chronicle (hereafter referred to as I.C.). Once *E* is safely in the Past, its full description is in the I.C. We may now think of the various parts of the I.C. as accounts to which practising historians endeavour to approximate their own accounts.

Let us say that every event in the Past now has its full description shelved somewhere in the historian's heaven. Remember: the events in the Past are 'fixed, *fait accompli*, and dead'. Only a modification in the events could force a modification in the I.C. But this is ruled out. The I.C. is then necessarily definitive. By contrast, the actual accounts offered to their audience by the working historians are always liable to modification. They may contain false sentences, they may have true sentences asserted in the wrong order, and they are almost certainly incomplete. At times bogus evidence or wrong interpretations of *bona fide* evidence may cause our historians to exchange a true sentence for a false one, so we shall want to distinguish a *correct* modification of an historical account. This, on our present view, will consist in bringing it into line with the I.C. Such a modification can then take at most three forms: (*a*) we add sentences which appear in the I.C. but not in the historian's account; (*b*) we eliminate sentences which appear in the historian's account but not in the I.C.; (*c*) we interchange the positions of all the remaining sentences in the historian's account to conform to the position of the corresponding sentences in the I.C. By repeated applications of these three rules of rectification we finally get a corrected version of the original account. It would in fact be an exact duplicate of the appropriate part of the I.C.

This is just the sort of thing a machine could do. Perhaps even the work of the Ideal Chronicler might be given over to a machine. The only place then where merely human effort is required is in the construction of an 'uncorrected account'. This, of course, has to be done through old-fashioned methods, e.g. gathering data, framing hypotheses, making and testing inferences, and the like. One is never sure of accounts which are constructed in this pedestrian manner: new evidence may turn up, a fresh hypothesis may be licensed by new scientific developments, completely new interpretations given when a genius appears. Painfully, old accounts are revised and replaced with new ones, and all the work that went into the earlier account has produced something now gone out of date. A thankless, endless business. What a pity it is the historian has not in his own archives a certified copy of the I.C. against which to check his own account by applying our few simple rules.

Well, let us *give* him the I.C.! Now he can know everything. Yet it is a pernicious gift. For what now is our historian to do? He can go into another field of history, but our bounty knows no end: we give him whatever parts of the I.C. he wants. Clearly there no longer seems anything for him to do *qua* historian, such as gathering data, framing hypotheses, constructing accounts, and so on. Why, after all, work hard to make shoddy accounts needing correction when the correct account is there to be read? To be sure, it may just have been in the use of the old practices that the historian's *raison d'être* was to be found. Sir Edmund Hillary would doubtless have taken it ill had a great hand reached down from heaven and set him atop Everest like a toy soldier. He would have arrived where he wanted to get, but nobody would recognize this as a great feat in mountaineering—not even if Sir Edmund had prayed for something like this to happen. For praying is not an exercise of the mountaineer's skill. I say: too bad for the historian. We shall have to remind him that history is not a sport, that his use of scholarly apparatus has always been a means to an end, namely the discovery of Truth. And this is just what we have given him. What is the difference if his historiographical tools turn out to have been *faute de mieux*? What more does he want or *can* he want?

Croce flings a similar challenge at those who see the task of history to describe the Past 'the way it really happened'. Suppose you have a complete description: what then will you do?[1] Croce says: 'Act!' I take this to mean: the historian must make some more history before he can write some more history, a distressingly sisyphean labour, something like a compulsive housekeeper who must keep scattering dust in order to go on fulfilling her essence. But I want to take this challenge to heart. What will be left for historians to do? They can, of course, simply be suspicious of the boon. Let them test it, then. It will always come out right if their methods are sound. Or they may take refuge in scepticism, but this will be just as damaging to ordinary historical practice as it would be to the I.C. Or they can ignore it. But is the historian to be like some Galahad who, turning the Grail about sadly in his hands, realizes that what he wanted after all was to just go on questing for it? There would be no point in this: further searching must henceforth be tainted with bad faith. The fly is in the fly bottle! The task of the philosopher is to lead it out.

My suggestion is: let him use the I.C. as he would any eye-witness account of an event in which he was interested. It will not tell him everything he wants to know about the event. This sounds as if it contradicts what we have said. Is not the I.C. definitively complete? And have I not said that nothing can happen to the Past to render it wrong or partial in any respect? Of course it is complete—but complete in the way in which a witness might describe it, even an Ideal Witness, capable of seeing all at once everything that happens, as it happens, the way it happens. *But this is not enough.* For there is a class of descriptions of any event under which the event cannot be witnessed,[1] and these descriptions are necessarily and systematically excluded from the I.C. The whole truth concerning an event can only be known after, and sometimes only *long* after the event itself has taken place, and this part of the story historians alone can tell. It is something even the best sort of witness cannot know. What we deliberately neglected to equip the Ideal Chronicler with was knowledge of the future.

Yeats, describing in his poem the rape of Leda by Zeus, writes: 'A shudder in the loins engenders there/ The broken wall, the burning roof and tower/ And Agamemnon dead.' Waiving for the moment questions regarding the historicity of episode, the *sentence* itself is of a kind which could not appear in the I.C. even if the event happened—in contrast with 'He holds her helpless breast upon his breast' which conceivably could appear there. For the latter describes what could be witnessed. But nobody could witness the act under the description 'Zeus engenders the death of Agamemnon'. For that king is not yet even born, and much will happen before his tragic end, as we now know. The death of

Agamemnon may be witnessed, only much later. Then someone might trace it all back to the violation of Leda, and he could see, in historical retrospect, that action of Zeus's as laden with a kind of destiny. To all of this the Ideal Witness is blind. Without referring to the future, without going beyond what can be said of what happens, as it happens, the way it happens, he could not even write, in 1618, 'The Thirty Years War begins now'—if that war was so-called because of its duration.

The class of descriptions I am concerned with refer to two distinct and time-separated events, E-1 and E-2. They *describe* the earliest of the events referred to. Yeats's sentence refers to the rape of Leda and to the death of Agamemnon, but it describes only the raping of Leda. 'The Thirty Years War began in 1618' refers to the beginning and to the end of the war, but it is about the beginning of the war. On the assumption that the war was so-called because of its length, nobody could presumably describe it in 1618—or at any time before 1648—as the 'Thirty Years War'. Of course someone might *predict* that the war would last just that long, and put sufficient confidence in his prediction to actually *describe* the war that way. But he would be making a claim on the future, which is what we are not allowing the I.C. to do. If we describe an event E-1 by making reference to a future event E-2 before E-2 occurs or is supposed to occur, we will have to withdraw the description, or reckon it false, if E-2 fails to happen. But the I.C. is so constructed as not to be mistaken at any point. There are to be no erasures. What it describes is fixed, and it says nothing which is not true. I shall later have more to say about predictions and descriptions, and I shall want, moreover, to explore some of the consequences of allowing the I.C. to make claims on the future. As matters now stand, however, it can make no such claims, and cannot, accordingly, employ the sorts of sentences— hereafter to be designated *narrative sentences*—I have just characterized. In this case there are no beginnings and endings in the I.C. 'If there are no beginnings and endings,' wrote Virginia Woolf in *The Waves*, 'there are no stories.' 'Cut away the future,' wrote Whitehead, 'and the present collapses, emptied of its proper content.'[1] It begins to dawn on one that a 'full description' does not adequately meet the needs of historians, and so fails to stand as the ideal which we hope our own accounts will approach; and that not being witness to the event is not so bad a thing if our interests are historical—which shows, I suppose, that some of the arguments of historical relativism are inappropriate.[1]

Factually false sentences may be converted into truths in two ways, provided the meanings of the words used remain constant: we may correct the sentences or rectify the facts they mean to describe. If there are three chairs in the room and someone says falsely 'There are four chairs in the room', he may achieve a true description by adding a chair or by striking out 'four' and replacing it with 'three'. With regard to false sentences about the past, however, I have only the option of correcting the sentences if truth be my aim. For some centuries there has been no opportunity to morally re-educate the Borgias so as to make the statement 'The Borgias were decent folk' come out true. At best I can replace 'decent' with 'wicked' or, if committed to the sentence, I can try to get the meaning of 'decent' changed—a self-defeating enterprise if I am committed to the proposition that the Borgias were *decent*. 'You can't make the Borgias decent' changes its meaning radically after 1503: before that time it might mean only that the Borgias were invincibly malfeasant, after that time that the appropriate Borgias, and the events in their lives, were totally embedded in the Past. Suppose, however, that there were a time-machine: our programme then might be to return to the Past, work hard on Alexander and his spawn, get them to walk in the ways of righteousness, and return to the present with a sentence made true *via* rectification of the facts. This is of course a hopeless undertaking, not because of the Borgias, but because of the unalterability of the Past. But *why* is the Past unalterable?

One may be tempted to say: because effects cannot temporally precede their causes, so the events of the Past cannot be the effect of causes now or at any future time in operation. Certainly the reason cannot *simply* be that the events in question are not 'here' so that we cannot, so to speak, lay hands on them: for future events are not 'here' either, and yet causes now in operation may be expected to have some effect on future events. On the other hand, the sort of situation I am considering differs from *this* one: a later event, say a coin falling heads, is said to cause an earlier event, say a man *saying* 'Heads'.[2] For in such a case, when the coin falls heads at *t-2*, the man has already actually said 'Heads' at *t-1*. But what would count as *changing* the Past would perhaps be something like this: someone undertakes to change the Borgias at *t-2*, the Borgias are vicious at *t-1*, the man succeeds in making them virtuous instead of vicious *at t-1*. To make the cases parallel, we should have to think of the man saying 'Tails' at *t-1*, the coin falling heads at *t-2*, and *this* then causing the man to say 'Heads' instead of 'Tails' at *t-1*.

Now if the Past cannot be changed in this manner, it cannot be simply because effects cannot precede their causes. For suppose the historian, interested in the latter day vindication of the Borgia's reputation, should admit that there is nothing *he* can do along these lines. Still, he might argue, they can change for all that. For there might be events *earlier* in the time-scale than the wicked behaviour of the Borgias which will still somehow cause the Borgias to mend their ways: it is simply that they have not yet discharged their causal energy, but have lain dormant all these centuries, like a volcano. This is surely an extravagant proposal, but the causes in question obviously precede their proposed effects, so the incapacity of the Past to change can no longer be charged to the temporal asymmetry of cause and effect. Moreover, we cannot simply say that the alleged events, earlier in the time-scale than the hoped for effects, must, just because they are past, be causally inoperative—for this would immediately entail a general argument against causality: our concept of causality requires action at a temporal distance. Otherwise no time-separated events can be related as cause and effect and we could not, accordingly, expect the future to be in any sense affected by things happening now. Worse, there would still remain the possibility that the events of the Past just spontaneously change, without anything causing them to do so.

But in the end all these difficulties are irrelevant. For what we are ruling out, so far as causality is concerned, is that any cause, earlier or later than an event *E* can act on *E* once *E* is past. For suppose *E* has occurred at *t-1*. Then any change in *E* will have to consist in either adding a property, eliminating a property, or both. Let *F* be a property to be added: then at *t-1* *E* is both *F* and not-*F*, which is contradictory by definition. But it would similarly be contradictory if a property *G* is eliminated: *E* would then be both *G* and not-*G* at *t-1*. This then takes care even of spontaneous change. But since E is at t-1, no change can take place in E at any other time, say t-2. For then something would have to be happening at t-1 and at t-2 at the same time, in other words, two distinct times would have to be simultaneous. And this again is contradictory.

When it comes to false descriptions of the events of the Past, then, the only means of converting them to truths is 'rectification of terms'. On the other hand, there is a sense in which we may speak of the Past as changing; that sense in which an event at t-1 acquires new properties not because we (or anything) causally operate on that event, nor because something goes on happening at t-1 after t-1 ceases, but because the event at t-1 comes to stand in different relationships to events that occur later. But this in effect means that the *description* of E-at-t-1 may become richer over time without the event itself exhibiting any sort of instability, and it is for this reason that what I have called the 'full description' of E at t-1 cannot be definitive.

Suppose that E-1 at t-1 is a necessary condition for E-2 at t-2. Then it immediately follows that E-2 at t-2 is a *sufficient* condition for E-1 at t-1.[1] A sufficient condition for an event may thus occur later in time than the event. We cannot readily assimilate the concept of cause to the concept of neccessary and sufficient conditions unless we are prepared to say that causes may succeed effects.[2] So it is difficult to suppose that E-2 *makes* E-1 happen. But at the very least it permits a *description* of E-1 under which E-1 could not have been witnessed and which, accordingly, could not have appeared in the I.C. Now there may be indefinitely many such descriptions, for each temporally later sufficient condition for E-1 affords a fresh description of that event. And precisely the same considerations apply for temporally later necessary conditions for E-1.

Suppose, for example, that a scientist S discovers a theory T at t-1. S perhaps does not publish T. At some later time t-2, a different scientist S' independently discovers T, which is now published and taken into the body of accepted scientific theories. Historians of science subsequently find out that S really hit on T before S'. This need take away no credit from S', but it allows us to say, not merely that S discovers T at t-1, but that S *anticipated* at t-1 the discovery by S' of T at t-2. This will indeed be a description of what S did at t-1, but it will be a description under which *S*'s behaviour could not have been witnessed and it will be an important fact about the event which accordingly fails to get mentioned by the I.C. Meanwhile, the historian who describes the event in this way will have used a *narrative sentence*.

For it to be true that a man *anticipates* T at t-1, it is *logically* necessary that T be later set forth, say at t-2. There are, however, some complications. We cannot *simply* say that the discovery by S' of T at t-2 was a necessary condition for the anticipation by S of T at t-1. We cannot, that is, simply say that had S' not hit upon T at t-2, S would not have anticipated T at t-1. For after all, some scientist other than S' could have arrived at the same theory, or S' himself might have discovered it at a different time than t-2. We can only say that for it to be true that S anticipates T at t-1, *someone*, at *some* time later than t-1, must also discover T. And obviously 'Someone discovers T later than S discovers T' is *not* equivalent to 'S' discovers T at t-2, and t-2 is later than the time at which S discovers T'. The former is entailed by, but does not entail the latter.

Nonetheless, a finer description of both events readily enough converts the latter into a necessary condition of the former. Let S be Aristarchus and S' be Copernicus. Then we might describe what Aristarchus accomplished at some time in 270 B.C. as follows 'Aristarchus anticipated in 270 B.C. the theory which Copernicus published in A.D. 1543'. If Copernicus had not published the theory, or had not published it at that time, or if someone other than Copernicus had published the theory at the stated time, this sentence about Aristarchus would be *false*. Hence, under the appropriate description, something done by Copernicus is a temporally later necessary condition for something done by Aristarchus. It immediately follows under just this description that what Aristarchus did in 270 B.C. is a *sufficient* condition for what Copernicus did some seventeen centuries later. It does not follow, of course, that what Aristarchus did caused, or figured as part of a cause of, the affirmation of heliocentrism by Copernicus. This would have to be established independently. In a way, of course, the concept of causality is not so clear as one would wish. What Aristarchus did may in no sense have caused Copernicus to discover the heliocentric theory, but in a very definite sense it caused Copernicus to *re-discover* the heliocentric theory.

It is not that Copernicus here did two distinct things: it was just the same action, seen under two distinct descriptions.

'Being a cause' may indeed be a special case of the sort of characterization of events which narrative description affords. Causes after all cannot be witnessed *as* causes: Hume pointed this out long ago. To say of E-1 that it caused E-2 is to give a description of E-1 by referring to another event (E-2) which stands as a necessary condition for E-1— under the appropriate description. If E-2 fails to occur, if it is false that 'E-2 takes place', then it would follow that 'E-1 caused E-2' is in turn false. From this it does *not* follow that E-1 is a sufficient condition for E-2. We would presumably not want to say in general that every cause of an event is a sufficient condition for that event. Nor again would we want necessarily to say that E-2 is a necessary condition for E-1. What it would be proper to say is that the occurrence of E-2 is a necessary condition for E-1 being a cause, or more precisely, a cause of E-2. Briefly, then, the occurrence of E-2 is not a necessary condition for the occurrence of E-1; it is only a necessary condition for E-1 being correctly describable as a cause of E-2; and accordingly the I.C. could not say, of E-1 when it occurs, that E-1 is a cause of E-2. Hence 'is a cause of' would not be a predicate accessible to the I.C.

Nor, as we have seen, would 'anticipates' be a predicate accessible to the Ideal Chronicler. But there are many more such examples. For it to be true that Petrarch opened the Renaissance, it is logically required that the Renaissance take place, though in point of fact the Renaissance might have taken place whether Petrarch opened it or not. Again, for it to be true that Piero da Vinci begat a universal genius, his offspring (in this case Leonardo) logically *had* to become a universal genius. Other examples would be: 'correctly predicted', 'instigated', 'began', 'preceded', 'gave rise to', and so on. Each of these terms, to be true of an event E-1, logically requires the occurrence of an event temporally later than E-1, and sentences making use of such terms in the obvious way will then be narrative sentences.

In addition to lacking narrative sentences altogether, the I.C. is deprived of certain referring devices; expressions which uniquely designate certain events, persons, and places, by making use of relative pronouns—'the place where ...', 'the person who ...',—where the blank is filled with an expression which refers to an event which takes place temporally later than the earliest time at which there is such an individual to refer to. Newton wrote his *Principia* from 1685 until 1687, when it was published. After that date it would be natural to refer to Newton as 'the man who wrote *Principia Mathematica*'. Indeed, from that time forward it would not be unnatural to refer to Newton by means of that expression no matter what period in Newton's life we were concerned to speak about. We may for that matter speak of Woole-thorpe as the place where Newton was born *or* the place where the author of *Principia* was born. We, but not the I.C., may say that the author of *Principia* was born at Woolethorpe on Christmas Day, 1642. The sentence 'The author of *Principia* is born in Woolethorpe' cannot appear in the I.C. for Christmas Day, 1642. Only after 1687 could this sentence, appropriately tensed, appear in historical writings.

The house in Woolethorpe still stands. It is the same house which peasants, or English yeomen, might have seen in the seventeenth century. It doubtless looks much the same now as it did then. We may make a pilgrimage there if we wish. We will see the same house which those yeomen and peasants saw. But we will see it as the birthspot and early dwelling of one of the greatest scientists of all ages, the place where Newton made those great discoveries in the Plague Year of 1665. Because of the importance of these discoveries, and hence the importance of the man himself, the house at Woolethorpe has for us a special sig-nificance. No one could have felt this significance in 1642: it is something which only events future to 1642 could bestow upon it. It is because of the significance we attach to *those* events, now of course in the Past, that we are sensitive to the significance of the stone cottage.[1]

We can visit the house at Woolethorpe, but we cannot visit it at the *time* when Newton was born: just to visit the Past would be to change the Past, and this cannot be. If *per impossible* we could witness the birth of Newton, we would see that event as fraught with a sort of destiny to which even the most ambitious mother must be blind. A shepherd on a hill in Greece might have seen a woman ravished by a swan (a monstrous enough occurrence), but he would not see engendered there the death of Agamemnon. This is something which could have been 'seen' only by someone who knew what could not have been known at the time. Were we able to visit the Past we would bring with us our knowledge of the Future (we would in effect be remembering events which occurred later in time than what we would be witnessing). We could only witness the Past as 'it actually happened' if we somehow could forget just the sort of information which may have motivated us to wish to make temporal journeys in reverse.

'But,' it might be argued, 'a clairvoyant might both witness a set of events as they happened *and* see them as significant in the light of future events. We, remembering Einstein's accomplishments, might have seen the old man in the light of these. Why might not one who *foresaw* these accomplishments see the young man in the light of these same achievements? Think of the Magi!' Well, perhaps. But we have not yet allowed the Ideal Chronicler precognitive gifts. He only knows what happens, as it happens, the way it happens. Every event is equally significant to him, or equally insignificant; which is to say that the category of significance fails to apply. How could it apply since he does not know the future? For it is only in the light of the future that the events he witnesses will take on a measure of significance.

If we refuse to allow the I.C. to make *any* claim on the future, to refer to future events, what language is it going to use to describe what happens, as it happens, the way it happens? I have argued that events cannot be described by the I.C. as causes, nor can it characterize them by means of narrative sentences. Narrative sentences refer to at least two time-separated events, and describe the earlier event. But in a sense this structure is also exhibited by a whole class of sentences normally used to describe *actions*. Is the I.C. then to be deprived of the entire language of action? I want to pursue this question, for it will help isolate some further features of narrative sentences.

Before the maiden voyage of the ill-starred ship *Andrea Doria*, a series of advertisements were run showing men painting pictures, carving statues, making mosaics, and the like. Under each such picture was printed 'This man is building a ship'. The pictures did not show men engaged in the sorts of skills obviously involved in the building of a ship, but we were to understand by this that the *Andrea Doria* was to be no ordinary ship. *If* we thought of such activities as mosaic-making as part of what would normally be done in the building of a ship, the advertisements would fail to make their point—a picture, of men laying a keel would not make the point that the ship in question was to be *extraordinary*. Yet if the expression 'building a ship' were incapable of being extended to cover such un-normal activities, the advertisements again would fail to communicate their message: we would be puzzled indeed if, under a picture of a man lying drunk in the gutter, it said 'This man is building a ship' in a way in which we are not at all puzzled by the pictures we were shown. The predicates of action obey extremely flexible rules: indefinitely many sorts of behaviour may be covered with 'is building a ship'.

Literally speaking, a man may just be putting a seed in a hole when we describe him as 'planting roses', or simply turning screws when we describe him as 'repairing the radio'. Yet no one expects such literal descriptions. We would no more think of correcting the description 'planting roses' by the more literal 'putting seeds in holes' than we would think of accusing a man of falsehood when he answers 'What are you doing?' with 'Planting roses' because what he is *literally* doing is answering our question. The range of behaviour covered by 'is planting roses' includes digging, fertilizing, sowing, even purchasing shovels and seeds, even reading seed catalogues or hiring expert gardeners. Indeed it is the rare case where the action-predicate is literally applicable, for instance, where a man is actually putting rose plants in the ground. The presence of roses is the *result* which all these separate pieces of behaviour are meant to lead to; and because we see some connection between them and such a result, we tend to describe these different pieces of behaviour in terms of the result. Let R be any result, and let E be any behaviour engaged in so as to bring about R. Then what a man is doing may *either* be described with E or R. Then 'a is R-ing' will be a correct description of what a is doing if a does E and E is a means to R. But in fact 'is R-ing' will generally cover a whole range of different pieces of behaviour $B_1 \ldots B_n$, so that when it is true that a is R-ing, we may provisionally suppose that $a B_i$'s, where B_i is a member of the range and where 'B_i's' is a *literal* description of what a does. The range marked out by a predicate like 'is R-ing' is almost certain to be very flexible, and of whomever it is true that he is R-ing it will generally be true that he will do different things in the range. Or it may be the case that 'is R-ing' is indifferently applicable to a group of individuals each doing *one* of the things in the range, such as, in a mass-production factory. I shall term predicates like 'is-R-ing' *project verbs*.

Now suppose *a* does B_i at *t*-1, and we describe his action with the appropriate project verb, '*a* is R-ing'. Is this not to be describing his behaviour in the light of some future occurrence, namely the coming about of *R* ? And does the sentence then not refer to two time-separated events, namely B_i at *t*-1 and *R* at *t*-2 ? But this would then seem to qualify all sentences which use project verbs in the way I have indicated as narrative sentences. Yet if we allow this, and if narrative sentences are ruled out for the I.C., it would follow that the I.C. could not use project-verbs, and the problem of how it would describe actions becomes intense. If, on the other hand, we permit use of project verbs by the I.C., are we not then allowing it to make claims on the future ? In which case why draw the line at all ? Or, if we decide that sentences employing project verbs are not narrative sentences, what further characterization of narrative sentences must we give in order to bring out the difference ? Let me take these queries up singly.

Suppose the I.C. were restricted to using only predicates of the sort which can appear in the range $B_1 \ldots B_n$ when ordinarily we would use project verbs. Then if we construe the relationship between terms in this range and project verbs as analogous to the relationship between phenomenal predicates and physical object terms, no difficulty need arise, at least in principle. For then a project verb would be eliminable in favour of a set of range-terms, and the I.C. would merely present a more detailed description of what people did than the use of project verbs affords. Such detailed accounts would be wholly consonant with what we expect of the I.C. We, in reading the I.C., if equipped with the appropriate rules of translations, could always replace a series of such descriptions with a single description using a project verb. Unfortunately, the problem of describing actions is even more complex than we have so far made it out to be. To begin with, it may be the case that a project verb is true of a man at a time when *no* term from the range $B_1 \ldots B_n$ is true of him. For a project verb may be true of a man during an indefinitely long period without his having, at every moment during that period, to do one or another of the things included in the corresponding range. We may speak of Jones writing a book all during the year. During that time Jones, amongst other things, sleeps. Yet the fact that he sleeps during that time does not falsify the claim that he is writing a book. Moreover, suppose a man does B_i and B_i is in a range marked out by 'is *R*-ing'. Still, it would not immediately follow that he is *R*-ing. Thus Jones may be digging holes, and though digging holes is part of what a man of whom 'planting roses' is true *does*, we could not infallibly infer that Jones is planting roses: he may be planting lilacs or *just* digging holes. But again, suppose Jones in temporal succession puts a seed in a hole, scratches his head, strikes a match, blows a smoke ring, thinks of his wife, and shifts his foot. Asked at any moment during this stretch of time, *what* he is doing, Jones will answer correctly 'Planting roses'. But only the first item in the series belongs in the range marked off by this project verb. In an important sense, then, reading the I.C. for this interval of Jones's morning will give no real notion of what Jones was doing unless we were able to collect from the series just those pieces of behaviour which, so to speak, belong together as part of the single project of rose-planting.

There is a kind of ambiguity in the word 'doing'. In one sense, if we knew all of a man's behaviour during a certain interval, we would *know* everything he was doing. In another sense, however, we should have only the raw materials for knowing what he was doing. In the one sense, the I.C. tells us everything we want to know, in another sense it doesn't. Not to have the use of project verbs is to lack the linguistic wherewithal for *organizing* the various statements of the I.C., but more importantly, for the I.C. to lack the use of project words is to render it incapable of describing what men are *doing*—and so disqualifies it from setting down whatever happens, as it happens, the way it happens.

Yet, if we do permit use of project verbs to the Ideal Chronicler, so that it can give a humanly coherent account of what goes on, have we not then violated our restriction against his making claims on the future? If the I.C. is to be allowed to say that Jones is planting roses, when he is only putting a shovel over his shoulders while setting off for the rose-fields, why can it not say that Mrs Newton is giving birth to the author of *Principia* when she has literally only been producing an infant with a weak neck? It would strike *anyone* as odd were he to be told that a universal genius had been born next door, though it would strike nobody as odd were he told that a rose had been planted next door—even though the rose could not be seen for some months. I venture to say that the difference lies in the sort of claim on the future which is made, and shall now try to make this clear.

When is a sentence like '*a* is planting roses' ever *falsified*? The question is exceedingly complex, due, amongst other things, to the indefiniteness of the range of things marked out by the project verb and to complications in the concept of intention. If we see a person just standing still, we cannot surely say that 'is planting roses' is false of him, even though he is at that moment engaged in no obvious activity at all: simply resting in the course of carrying out his project. Nor, if we *ask* him what he is doing and he sincerely replies 'planting lilacs' does this falsify the proposition that he is planting roses, for even though he does not intend to plant roses, he is in fact doing just that, having mistakenly assumed that the seeds were lilac when in fact they were rose seeds. *If* lilacs rather than roses come forth, this will perhaps falsify the proposition that he was planting roses, provided that we are certain that no one surreptitiously replaced his rose seeds with lilac seeds. But if roses fail to come forth, this does *not* falsify our proposition, so long as he did *whatever* might, by current criteria of rosiculture, count as planting roses. So let us assume that there is a definite range of operations, the doing of which constitutes planting roses, and let us suppose further that these operations constitute necessary conditions for the coming forth of roses (forgetting about wild roses). If this were the case, then failure to do these things would not merely guarantee the non-coming-forth of roses ('these things' being necessary conditions for that) but would also falsify the claim that the person was planting roses. On the other hand, since the operations are merely necessary conditions, should *a* do all of them there would be no guarantee that roses would come forth—a hurricane might come up and undo all *a*'s labours—but it *would* be true that *a* was planting roses.

So it can be the case that, while true that *a* is planting roses, it will be false that roses come forth. More generally, if 'is *R*-ing' is any project verb, it may be the case that a man is *R*-ing without it having to be the case that *R* happens—where *R* is the accepted outcome of *R*-ing. A man then may correctly be said to be repairing the radio—though the radio fails to get repaired—providing only that by common criteria the man is doing the things which fall within the admittedly elastic range marked out by 'repairing the radio'. Hence, though a sentence which asserts a project verb of someone may indeed refer to two time-separated events —B_i which the man *literally* does, and *R*, which is the expected result— and describes the earlier event in the light of the later one, it is not *logically* required that the later event take place for the sentence to be true. So, when we correctly say that *a* is *R*-ing, the reference made to the future does not enter as part of the truth conditions for the sentence.[1] Accordingly, the I.C. might be allowed to say that *a* is *R*-ing without making the kind of claim on the future which would require an erasure in case *R* fails to result. So *R* is not what we earlier called a 'future-referring' term.

Now Jones, sowing rose seeds, is planting roses come what may. It may turn out that he has planted roses which come forth and win prizes at the rose festival. This would allow the *narrative* description, covering exactly the same actions that 'Jones is planting roses' once covered, that Jones was planting prize-winning roses. Two witnesses to Jones's actions might say, respectively, 'Jones is planting roses' and 'Jones is planting prize-winning roses'. The first will be right no matter what the future brings. The second will be wrong if the future brings no prizes to Jones's roses, or if indeed no roses of Jones's come forth. Unless the second man is merely expressing his hopes or saying encouraging things to Jones, his sentence is exposed to more exacting truth conditions than is the first man's. For *his* sentence to be true it is *logically* required that Jones's work result in roses, and that the roses result in prizes. In this sense, he is making a stronger claim on the future than the simple 'Jones is planting roses' does.

In the past tense, 'Jones was planting the prize-winning roses' does, and 'Jones was planting roses' does *not*, require, for its truth, the resultant coming forth of roses. A narrative sentence then does not merely refer to two time-separated events and describe the earlier with reference to the later. It in addition logically requires, if it is to be true, the occurrence of *both* events. In the present tense, 'Jones is planting roses' is not, while 'Jones is planting the prize-winning roses' *is* partially predictive. As a prediction, it will have been false if no roses come forth (and if they fail to win prizes). Were the I.C. then to have said 'Jones is planting the prize-winning roses', an erasure would be called for unless the later event come out. To guarantee no erasures, we must either prohibit use of narrative sentences in the present tense, or grant special cognitive powers to the Ideal Chronicler. Before considering *that* alternative, I want to introduce some further complications.

I have claimed that a project verb may be true of an individual through an extended spell of time without the individual needing to be doing, at every moment during that time, one or another of the specific actions in the range marked out by that project verb. This follows from the fact that more than one project-verb may be true of an individual during the same temporal stretch: *a* may be writing a book and courting a widow all during June. Suppose we are interested not in *a*'s total biography, but merely in the history of his book. Then we shall require some criteria for picking out all and only those performances of *a*'s which are exercises of his authorship or which are related in some manner or other to these. Which events in *a*'s life we shall thus collect will depend very much on our criteria for what counts as writing books: the extent of our collection will vary with the stringency of our criteria. Moreover, *a* is almost certain to be engaged with other projects during that time, so there will be gaps between the events our criteria enable us to collect. The events we *do* collect will constitute a gerrymandered sub-set of whatever *a* does during the time covered. '*R*-ing' is continuously true of *a* so long as *R*-ing is his project, but *a R*'s only intermittently through that time.

Inasmuch as we have adopted the convention of regarding events as extended in time, projects are time-extended events. But given the chequered history of typical projects, we may classify events as continuous and discontinuous, roughly on the analogy of the distinction between smooth and dotted lines. A dotted line is a series of smooth lines with separating interstices, and a discontinuous event may then be characterized as a series of continuous events separated by irrelevant happenings. True, on a microscopic inspection what looks smooth to the naked eye will be shown up as riddled with breaks. So in the end the difference may be one of degree only, and I have no wish to argue, by transcendental deduction, so to speak, that there must be ultimate, smooth lines. No more do I wish to argue that there must be continuous events if we set our temporal termini sufficiently close. Indeed it is much more to my point that there should be discontinuous events in the sense illustrated by the history of a's book. The difference I mean to bring out is essentially that of a project and the serial events which count as in the range of actions marked out by the use of the appropriate project word. Briefly, if B_i and B_j are in the range of 'R-ing', then if B_i is done at t-1 and B_j is done at t-1 plus delta-t, and if nothing is done in the interval between B_i and B_j which is in the range of 'R-ing', R-ing will be discontinuous and each of B_i and B_j will be continuous relative to R-ing. Events discontinuous in this sense I shall designate as *temporal structures*.

Now such projects as writing books and courting widows are amongst the simpler sorts of temporal structures. Some projects, for example, involve numbers of individuals. With some violence to ordinary usage, we may speak of innumerable Frenchmen as engaged in French-revolutionizing during an interval of time in the neighbourhood of 1789. The makeshift project verb 'is French-revolutionizing' is not, of course, true of every individual in France during that interval, and is true of some individuals not in France. Nor, of those of whom it is true, were each of them at every moment during that interval French-revolution-izing. So not everything which went on in France is within the range marked out by our project-word: the project was exhibited then discontinuously over French soil and eighteenth-century time. Just which happenings there and then are to be counted part of the temporal structure denoted by 'The French Revolution' depends very much on our criteria of relevance. Doubtless there are shared criteria so that no disagreement exists over certain events. But insofar as there is disagree-ment over criteria, the disputants will collect different events and chart the temporal structure differently, and obviously our criteria will be modified in the light of new sociological and psychological insights. The Past does not change, perhaps, but our manner of organizing it does. To return to our map-making metaphor of Section II: there is a sense in which the territories (read: temporal structures) which historians endeavour to map *do* change. They change as our criteria change, and at best our criteria are apt to be flexible, as we saw when speaking of ship-building.

Any term which can sensibly be taken as a value for x in the expression 'the history of x' designates a temporal structure. Our criteria for identifying a, if a be a value of x, determines which events are to be mentioned in our history. Not to have a criterion for picking out some happenings as relevant and others as irrelevant is simply not be in a position to write history at all.[1] Temporal structures are, of course, *ad hoc* in some degree. The identical event may indeed be a constituent in any number of different temporal structures: E may be collected with any number of otherwise disjointed collections of events into distinct temporal wholes. Thus, our description of E may accordingly vary as we group it with different collections of events into different temporal structures. Thus to describe E with a narrative sentence—to relate it to some later event E'—is to locate both E and E' in the same temporal structure. But no *a priori* limit may be set to the number of different narrative sentences, each of which truly describes E, and hence no limit may be set to the number of different temporal structures within which historical organization of the Past will locate E.

Nonetheless, just as different contexts will determine which of the innumerable possible descriptions of an object is the appropriate description to give, so the particular temporal structure in which an historian is interested will often determine which is the correct description of a given event. I have contended that a particular thing or occurrence acquires historical *significance* in virtue of its relations to some other thing or occurrence in which we happen to have some special interest, or to which we attach some importance, for whatever reason. Narrative sentences then are frequently used to justify the *mention*, in a narrative, of some thing or event whose significance might otherwise escape a reader. A novelist, for instance, may interrupt his story in order to comment narratively on some happening to which he wants to draw our attention, for example, 'Little did Smith know that his innocent sally was to cause the Bishop's death'. He thus refers ahead to that particular episode from which the earlier and otherwise trivial-seeming event derives its import. Historians, too, often use such devices. Why, in a history of the Crimean War, single out Captain Nolan for special mention when so many soldiers are not spoken of? Because, when Captain Nolan joined Lord Raglan's staff, 'It was a fatal moment'.[1] 'This officer, brave, brilliant, devoted, was destined to be the instrument which sent the Light Brigade to its doom.'[2]

Words like 'fatal', 'detained', 'doom', dramatize what is an essential fact about the historical organization of the Past. The Charge of the Light Brigade was a piece of idiotic splendour that impressed the minds of men: it was a fit subject even for poetic treatment. Had it never taken place, or had it been routine or inglorious, the light of historical interest might never have fallen on Captain Nolan, or might have illuminated him differently, for example, in some other temporal structure, say the history of cavalry.

Examples of such retroactive re-alignment of the Past might be multiplied indefinitely. Any novel philosophical insight, for instance, may force a fresh restructuring of the whole history of philosophy; one begins to see earlier philosophers as predecessors—which, ironically, can lead men to understress the originality of him whose novel insight brought to historical attention otherwise unremarked traits of antecedent philosophical utterances. Kant complained bitterly about this.[3] We have recently seen, as a result of the products of the New York School of abstract expressionism, a comparable revaluation of Monet. One might find that Monet influenced not a single member of the New York School: but because these men began to paint in a special way Monet *became* a predecessor in his late works. 'If,' wrote Bergson, 'there had been no Rousseau, no Chateaubriand, no Vigny, no Victor Hugo, not only would one never perceive, but indeed there would *not have been* any romanticism in the classics of the past.' For

this romanticism of the classicists was only actualized by the carving out of a certain aspect of their work. But this *découpure*, with its specific form, no more existed in the literature of classicism before the advent of romanticism than the amusing design exists in a passing cloud before an artist perceives it there in organizing that formless mass according to his fancy.[4]

This, of course, is extravagantly put. I should prefer to say that the romantic elements were there, in classicism, to be discovered. But it is a discovery for which we require the *concept* of romanticism, and criteria for identifying the romantic. But a concept of romanticism would naturally not have been available in the heyday of classicism. I want parenthetically to remark that whatever in classical writings turns out to fall under the concept of romanticism was doubtless put in those works intentionally. But they were not intentional under the description 'putting in romantic elements', for the authors lacked that concept. This is an important limitation on the use of *Verstehen*. It was not an *intention* of Aristarchus to anticipate Copernicus, nor of Petrarch to open the Renaissance. To give such descriptions requires concepts which were only available at a later time. From this it follows that even having access to the minds of the men whose action he describes will not enable the Ideal Chronicler to appreciate the significance of those actions.

To be alive to the historical significance of events as they happen, one has to know to which later events these will be related, in narrative sentences, by historians of the future. It will then not be enough simply to be able to predict future events. It will be necessary to know *which* future events are relevant, and this requires predicting the *interests* of future historians. I want now to turn to the matter of predicting events in this manner. But I note in passing that if the Ideal Chronicler is to do this, it will be the works of human historians which will be his models rather than, as we earlier supposed, the other way around.

We cannot identify a sentence S as a prediction simply by tense, for some sentences may be predictions and yet atypically be in the *past* tense. Thus 'Aristarchus anticipated Copernicus' is predictive at any time after 270 B.C. and before A.D. 1453.[1] Nor is it simply a matter of the user of S intending S to be a prediction, for the user may be confused on dates, and the race whose outcome he *tries* to predict may already have been run and won by the time he utters S.[2] I shall stipulate, not a definition of, but only a necessary condition for, predictive sentences: S is a prediction when S refers to E and E does not occur earlier than, or concurrently with, the utterance of S.[3]

A narrative sentence, referring as it does to a time-ordered pair of events E-1 and E-2, will then be a prediction if used by the Ideal Chronicler. For he will write it down *when* E-1 takes place (narrative sentences being about the earliest of the events they refer to), and so temporally earlier than E-2. Moreover, if the I.C. is to remain definitive, these must be *correct* predictions. But this now modifies the task of the Ideal Chronicler considerably. For since the pair of events referred to by a narrative sentence belongs to the same temporal structure, the Ideal Chronicler has to be structuring the Future in the same way that future historians will be structuring the Past. Since the I.C. is to be *complete*, all narrative sentences true of E-1 must be written down at once, and accordingly the Ideal Chronicler must lay out all the temporal structures in which E-1 will be located. In effect the I.C. is writing history before it has happened. So if we *now* allow pieces of the I.C. to fall into the hands of historians, they will find out a great deal more than simply what happened, as it happened, the way it happened. They will also find out what *will* happen (unless the events, whose account they have, are totally unrelated to future happenings). But with this we destroy the asymmetry in our concept of Past and Future: Past and Future now are one in point of determinateness. Indeed, this is analytic. For the truth of p is logically entailed by the truth of 'a correctly predicts that p', and every prediction made by the Ideal Chronicler is by definition correct.

So everything is changed. In particular the cognitive powers of the Ideal Chronicler have been changed. Before, though he was privy to a great deal more than a mere human could be, his manner of knowing was simply an extension of a familiar human cognitive situation: he *witnessed* the events he wrote about. But one cannot witness *future* events without changing the meaning of 'witnessing'. How then *can* he know the future? Is the behaviour of the Ideal Chronicler any longer even intelligible to us? Let us turn to more strictly human cases in which predictions are made, and work our way gradually back to these questions.

When, at t-1, a man predicts E-at-t-2, we may always ask how he knows, or why he thinks, that E-at-t-2. This will generally be by way of request for evidence, and our confidence in the prediction will vary with our assessment of that evidence. Let the prediction be 'Rain at t-2'. Then evidence may range from rheumatic twinges or mere hunches, to gravid clouds or the behaviour of birds, to the outcome of tests with cloud chambers, X-rays, electronic diffraction, or the like. Or it may simply be the weather report in the paper. Whatever the case, that which is cited as evidence is accepted as such only when some answer can be given to the question *why* it is thought to provide some basis for believing that rain at t-2. The answer may range from a plain inductive generalization to the latest meteorological theory. Briefly, we need, for predictions, some event and some law-like sentence or other which allows us to infer, from that event, a future happening. Now I am for the moment not interested in whether something is good or bad evidence, but only in the most general requirement which that something must satisfy if it is to be evidence *at all*, namely, that whatever is offered in evidence must be available at the time the prediction is made. Given our characterization of predictions, *one* thing systematically excluded by this requirement is the event predicted. Any statement to the effect that E will happen, when E has already happened, will automatically be false by virtue of misrepresenting the temporal relationship between the utterance of that statement and E. Hence E, if offered as evidence for a prediction about itself, will automatically render that prediction false.

At t-2, then, we have access to information *in principle* unavailable to a man who predicted what would happen at t-2. Specifically, we are in a position to *know* that his prediction was correct or incorrect. Asked how we know that it is raining, we can in principle show evidence which even the most sophisticated forecaster could not have shown earlier: we can point to the rain-fall. Now if narrative sentences refer to two time-separated events, and are predictive until the second event occurs, it would seem that after that event, persons (historians) can always cite evidence in favour of the narrative sentence which would in principle have been unavailable before the occurrence of the temporally latest event referred to by it: they can cite the *event itself*. And they are then in a position to know, as no one before the occurrence of that event would be, that the narrative sentence is true. Whether it was true before is a question for our next chapter: I am interested only in the epistemology of the matter here.

But if we are really doing epistemology, we have leapt too far. For suppose that at *t*-1 it is predicted that E-at-*t*-2. Then indeed someone will, at *t*-2, have information lacking at *t*-1, namely the event itself, if the prediction turns out to have been correct. Presumably he witnesses E, whereas only signs of E could be witnessed at *t*-1. But then E can be witnessed *only* at *t*-2: at *t*-3 it is already too late for that, and so on for every *t*-*n* (*n* = 3). From *t*-3 on we are roughly in the same position as he who predicted at *t*-1: like him, we can only witness *signs* of E-at-*t*-2. In a sense, we are in a less favoured position. For the predictor may at least hope to witness the event he has predicted. But our own argument systematically falsifies 'I will witness E' if E occurs temporally earlier than the utterances of this sentence. The predictor is in a position to *be* in a position of witnessing, and hence knowing whether or not he predicted correctly. But not the retrodictor.

This disadvantage is partially offset by the fact that those who predict the occurrence of E and those who retrodict the occurrence of E may witness disjoint classes of signs of E. Possibly wet streets are not stronger signs that it has rained than gravid clouds are that it will rain, but copies of *De Revolutionibus Orbium Caelestium* would appear offhand to be stronger signs that *somebody* wrote the book than any signs I can think of would be that somebody will write it. At any rate, the retrodictor may have the testimony of witnesses to an event, and this sort of evidence is systematically outlawed for him who predicts, given our general restriction. As a special case we have histories of events after and not before they occur.

Granted that the possibility of someone saying he has witnessed an event, and is now waiting for it to happen, is ruled out by our restriction; granted again that we would simply find absurd the statement by somebody that his book has been published so that he had better get busy and write it, is it similarly to be granted as absurd that someone should claim to have the history of a set of events written, it only now being a matter for the events to *happen*? Let us try to imagine such a case.

Suppose we pick up a book titled *The Battle of Iwo Jima*. It describes in *minute detail* the men and movements of that conflict: it tells who gets wounded and when, who gets killed and why, and we then discover that the book was written in 1815! For all that, we find that the book tells us more than we already know, even if we are, say, the foremost historical experts on that battle. Using the book as a guide, we look up survivors heretofore unknown to us. Their testimony always squares with this strange, anachronistic windfall, which now becomes an invaluable guide to historical research, like a treasure map!

After all, a man may draw the treasure map *first*, and then afterwards place the treasure, or have the treasure placed. A man may lay out a programme and afterwards carry it out, or have it carried out. Here are instances of 'rectifying the facts'.[1] Why then cannot we write a history before the events about which it is written actually happen? Someone might argue that we would not *call* this history, that history by definition is about the past, that it violates usage, accordingly, to say the history of happenings in 1945 can be written in 1815. I am not one to quibble over usage: let it not be called 'history'. But suppose we only found out *after* having accepted the book as the definitive account of the Battle of Iwo Jima that it was written in 1815. I should find cold comfort in the fact that we could no longer call it history. It is the possibility of such an account, however it be designated, that I find disturbing.

A baby in the course of its babbling might, by sheer accident, utter a string of vocables which turn out to be a proof of Fermat's last theorem. Call this a coincidence: one string of vocables may be equiprobable with any other string. Or consider the baby an oracle, and bring in mathematicians to heed its noises. Anything counts as reasonable in such a case. But suppose our problem manuscript is discovered in a bundle of papers, the literary estate of a nineteenth-century writer, and there are with it letters. Typical of what these say may be: 'I have been hard at work on my book on Iwo Jima. The work goes slowly. ...' Enough such secondary documentation convinces us that the book is due to deliberate human contrivance. We find passages crossed out and replaced with what turn out to be factually correct emendations, all in that quaint nineteenth-century handwriting. Everyone would say: this is a forgery. But if we found amongst Newton's papers a celestial map for the year 1960, and checked and found it wholly accurate, we would not suspect a hoax. We would not feel the unease which comes when a fundamental concept is threatened. Why though?

Wittgenstein wrote: 'The future is hidden from us. But does an astronomer think like this when he calculates an eclipse of the sun?'[2]

The question is rhetorical: astronomers apparently *don't* think that way. The thing is, we know more or less what the astronomer does: he determines initial positions, solves equations, and so on. Our precocious historian wrote: 'The work goes slowly.' But what *kind* of work? And here we *don't* know. We only know that it can be not at all like what historians commonly do: work through the archives, authenticate documents, sift testimony, interview survivors and examine photographs. Our inclination now may be to say there can be no writing of history before events because nothing is to count as historiography. For the astronomer, the future is no more hidden than the past, and prediction and retrodiction are of a piece. But there is a special asymmetry between signs and traces of events, which we have already noticed. Footprints exist after, not before footfalls. Photographs, eye-witness reports, and the like exist after, not before the events they testify to, and it is with such things that historiography has to do. Think of the immense difficulties in trying to predict just those spots upon which a man's feet will fall as he crosses the sand: and how simple it is, so long as footprints remain, to retrodict the positions.

These asymmetries run deep. Seeing gravid clouds I may say 'It will rain unless ...' and seeing wet streets I may say 'It rained unless ...'. But the expression which will indifferently complete either sentence is rare. Thus 'a water-truck came by' naturally completes the second sentence, but—shifting tense—'a water-truck comes by' fits ill with the first. Again, 'It will rain unless the wind shifts' may be said when gravid clouds are seen, but 'It has rained unless the wind shifted' sounds odd when wet surfaces are seen. Moreover, if a man witnesses E at t-2, he is still regarded as a witness at t-3 but, although he will witness the event, he is not regarded a witness at t-1.

Yet, if we use the testimony of such a witness as the basis for a retrodiction, we are relying on his memory. Why might there not after all be a symmetry with using some precognitive deliverance from he who *will* be a witness as the basis for a prediction? Call such a person a 'pre-witness'. A pre-witness precognizes that he will witness, the way a witness remembers that he *has* witnessed an event. Someone might now argue: to say that a is a pre-witness is logically to presuppose that a will witness E, and to say that a will witness E is logically to presuppose that E will occur. But we cannot then accept, as evidence that E will occur, a's testimony—as a pre-witness—to that effect, for to *accept* him as a pre-witness is logically to presuppose the very question in issue, namely, the occurrence of E. Unfortunately, an exactly analogous argument would disqualify the evidence of witnesses, for to accept b as a witness of E logically presupposes that b did witness E. This in turn logically presupposes the occurrence of E. Hence to accept b as a witness, and his testimony as evidence for the occurrence of E, is to beg the question in issue. The truth of p is entailed by the truth of 'b remembers that p'. But then the truth of p is also entailed by the truth of 'a precognizes that p'.

Of course, if we insist on regarding precognition as symmetrical with memory, we would presumably have to rule out precognition as that upon which the 'historian' of the Battle of Iwo Jima based his account. For if we cannot remember events we have not witnessed, we cannot precognize events we *won't* witness, and the 'historian' will almost certainly not witness the battle. So the alleged symmetry between memory and precognitions comes to very little. This hardly affects the typical historian, who seldom has personally witnessed the events he writes of: but it is disastrous for the person who writes of events he *will* not witness.

Perhaps then he has some sort of second sight, and bases his account on prophetic visions. We might then explain his emendations on the grounds that a later vision supersedes an earlier one, as in the composition of the *Koran*. Yet we might ask how he really knows he has this sort of second sight, how he distinguishes between having a proper vision and simply imagining things. It may be that what he meant by 'The work goes slowly' was: 'Visions are few and far between.' But how should he distinguish his case from that of a novelist with a grudging muse? Note that we can match our strange prophet with an equally strange person who has *retroactive* visions: a person who writes, in 1960 and on the basis solely of visions, the history of what happened in 1815! Suppose indeed that this person writes in this manner a wholly accurate account. But at least we can *check* this man's visions against standard accounts. Even when he reports things not in standard accounts we can know in principle what sort of evidence it would take to verify what he says. But in 1815 there would have been nothing comparable against which the 'History of the Battle of Iwo Jima' could have been checked. Certainly not against *other* accounts. For the question would then arise how these accounts were arrived at. If they too were written on the basis of visions, we would only have transferred the problem. A visionary history and an orthodox history might arrive at the same conclusions: there would be orthodox ways of checking both. But when the account is written *before* the events in question, there are neither orthodox accounts nor orthodox ways of checking unorthodox ones. There may *be* such visions. Having them is just someone's huge good luck—like begetting a universal genius. Piero da Vinci's behaviour is instructive: he tried to duplicate the exact circumstances under which Leonardo was conceived in the expectation of duplicating Leonardo. You can say he did just the right thing, or just the wrong thing. It makes no difference. For in the end nothing is right or wrong when it comes to begetting the universal genius. There is no recipe.

However, when the astronomer calculates the future eclipse, we don't suppose *him* to enjoy special precognitive gifts or to require second sight. When we say the Future is hidden, all we may mean is that we lack the sorts of laws and theories which the astronomer *has*. Might not the precocious historian have used Science? By 'The work goes slowly' we are now to understand that he meant: it is damned hard to determine values for all the variables, damned hard to make those intricate calculations which lead deductively to the conclusions presented in 'The History of the Battle of Iwo Jima'. Well, this may very well be. We have fair reasons for believing that there were no such theories in 1815. We lack them today. And we cannot then really understand, since we don't ourselves have such theories, what sorts of things counted as initial and boundary conditions. But let us suppose the man knew these things, and that his work was 'scientific' work. He predicted the Battle the way the astronomer predicts the eclipse.

Once again, let us work from simple cases. We shall suppose a theory T in accordance with which an event E may be predicted from another event C. Let T be: 'Whenever gravid clouds, then rain.' The vocabulary of T then consists of two special terms, 'gravid clouds' and 'rain'. Now many things are true of rain-storms other than their just *being* rain-storms. Accordingly we may readily frame a description *D* of *E* which cannot be formulated in the skimpy lexicon of *T*.

Now *E* may certainly be predicted by means of *T*, but not under description *D*. In order to be able to do *that*, we shall have to show that the predicates of *D* are explicitly definable with the terms already included in *T* or—more likely in the present case—we shall have suitably to enrich our stock of terms. *T* becomes proportionately more complicated as a consequence of this, and we shall now suppose *T* to be brought to that level of complexity currently exhibited by the latest theory of meteorology. Supposing the vocabulary of *T* now consists of a set of terms F_1, F_2, F_3 ... F_n, we may say that the description under which *E* is predicted will ideally use each of these terms or its negate. This will then be the fullest description current theory affords.

We all know, of course, that any such description, however rich, is meagre in contrast with what is logically possible: that every predicate in the language (or its negate) could apply to *E*, and that even then, since *individuum est ineffabile*, the properties of *E* would not be exhausted: the richness of *E*'s properties outdistances the maximum richness of descriptive power in our language taken *in toto*. But this does not concern me particularly. For suppose *E* has happened, according to its prediction. Then there may be descriptions of *E* which we find it important to give, but which fall outside the linguistic reach of *T*. It may not have been just a rainstorm: it may have been the rainstorm in which our basement was flooded, or which washed away the wharf that Smith built in 1912. I do not mean to say that these things could not have been predicted. I only mean to say that they could not have been predicted by means of *T* alone. For 'floods Jones's basement' or 'washes away Smith's wharf' are almost certainly not terms true of rainstorms which are included in *T* or explicitly definable by means of the terms which are.

It is generally admitted that a scientific theory cannot predict an event under every true description of that event. Indeed, part of what we think of as scientific activity consists in finding the appropriate language for describing events, picking out those terms which designate relevant properties of events, or making up terms for this purpose. It is quite enough to know the initial position and motion of a body to be able to predict its path: one need not also know that a particular such body is a china egg made for the elder daughter of Czar Nicholas. So it would be pointless and, in the end, destructive of the whole concept of a scientific theory to recommend incorporation into a theory such as T those terms with which our own local interest in basements and wharves move us to describe rainstorms. Moreover, it would be an impossible demand. For there is no end to the number of temporal structures in which the historians of the future might see E as located. It may come to be known as the storm in which Alice and Bernard had their fatal quarrel, or during which the man who solved Fermat's last theorem was born. So it is a sufficient achievement to be able to predict E under *some* description of it. The claim, less frequently met with today than formerly, that there are two distinct kinds of events—scientific events, which can be predicted and explained, and historical events, which cannot—is erroneous. There are not two classes of events, but perhaps two classes of descriptions. Science may indeed fail to give us the information about events which we want, but this is because such information cannot always be stated in the abbreviated language of scientific theories. It would *destroy* the concept of meteorology to make such demands.

So it may be, but we are now interested in a different theory: the one used to predict, not just the occurrence of the Battle of Iwo Jima, but that event under the enormously detailed description found in our controversial 'history'. There must be such sentences in the latter as: 'At 3.30, February 20, Sgt Mallory, while arming a grenade, is killed by Pvt Kito—the latter's fifth and only successful shot of that day.' Small wonder the work went slowly! It would be labour enough merely to write the *history* in such detail. At any rate, the theory used to predict all this must be as linguistically rich as ordinary language. We are supposing after all that the account is readily intelligible to the plain reader.

But then suppose the manuscript had been discovered in, say, 1890. Readers then might have been puzzled by some of the language (as we are often puzzled by some of theirs), but, struck with the fertility of the writer's imagination, they might have assigned it to the same genre as the writings of Jules Verne—though it would perhaps be too wordy, too detailed, for a proper novel. Edited versions, even boys' editions of it would appear. Only after 1945 would people see it as pre-written history. Or suppose it had been discovered sometime in 1944 and really taken seriously as a piece of scientific prediction. The High Command might discuss it, compare it with their own plans, perhaps even alter their own plans. Sgt Mallory would see to it that he was elsewhere at 3.30 on 20 February. And then all the work, which went so slowly, would have come to naught: the predictions were false! For men refused to follow the manuscript, behaving like rebellious actors dissatisfied with the script. It is a common enough thing to falsify predictions. Someone predicts that the ball will strike the ground at a certain time, and someone else catches it. Surely it would be to a man's interest to falsify the prediction that he will lose his life at a certain time and place. The only way for the prediction to go through *is* for it to be discovered after the event. For we cannot, remember, change the Past.

Maybe that man in 1815 was aware of all this. Maybe he even so predicted the future that the manuscript would fall into peoples' hands in 1944, and that they would try to falsify the predictions made in it. He predicted what they would do, and wrote about this! Then the same situation as before would arise if this 'fuller' account were to fall into men's hands in 1944. What we cannot imagine is their knowing what prediction was made and *not* being able to falsify it, so long as the event predicted had not already happened. Imagine having the prediction that one will move ones left foot at *t*-1 and ones right foot at *t*-2. One tries to falsify this: one tries just to stand still at *t*-1, or move one's right foot, but *despite* all ones efforts, the prediction comes out! The feet fall into foreshadowed foot-prints: as though one had lost all control of one's limbs, they are now moving off on their own. Or imagine trying not to yell, and then finding despite this a scream tearing past ones lips. Think of a whole battlefield of men going through this weird alienation. In horror, men find themselves aiming guns; fingers move spontaneously to grenades and pluck out pins; men try to shout 'Retreat' but the predicted 'Attack!' comes out instead. Everyone watches his own behaviour in an almost spectatorial way, detached from every act, knowing in advance what will be done and unable to *do* anything to prevent its happening. These things happen in nightmares perhaps, or in the dreams of the Mad Scientist. In dreams it might happen that someone shouts 'Stop falling!' and I, in my flight through space, obey —arresting myself in mid-air. 'Stop falling!', in a real context, is a paradigm case of an order we cannot obey. 'Move your right foot!' in normal contexts is a paradigm case of an order we can disobey if we wish. The elaborate case I have just imagined could only occur if men lost what we normally regard as control over their actions. The one book we cannot imagine the man of Iwo Jima having in their hands is 'The History of the Battle of Iwo Jima'. Or rather: we cannot both imagine their having it *and* the book remaining true.

What we don't know, then, is what the historians of the future are going to say about us. If we *did*, we could falsify their accounts in just the same way that we could falsify predictions made before the time at which we are acting, or we can do this within the limits of normal human control; a set of limits which we expect science to widen rather than narrow.

So let us now suppose that the Battle was predicted, and the prediction then only discovered afterwards. We regard it as a great achievement, and regret only that it was discovered too late. Since it was discovered too late, it is true. Nothing can happen to the Past to make it false, but as time passes, we will find it more and more necessary to add fresh descriptions of the Battle of Iwo Jima. A man who was a private then survives, due to the heroic action of a man whose dying thought might have been that he had sacrificed himself for so insignificant a person. That private then goes on to do great deeds! The episode takes on a special significance: it is taught to school-children. They enjoy enacting the scene in which was saved the life of the man who . . . . And more and more narrative sentences enter later accounts of the Battle: sentences which even the genius of 1815 did not know.

Could the Ideal Chronicler have known? It is up to us to say. He is our creation, we can do with him as we will. It was we, after all, who *decided* that he should be capable of simultaneously transcribing everything as it happened, when it happened, the way it happened. But why prolong the fiction? It has served our purposes, and may now be abandoned. And with it goes the I.C., of which we failed to find a version which did not either tell us less than we want to, or more than we can know. What about our lame metaphysical model? What did it serve to do except to say metaphorically that true sentences about the Past are not false—which may be all that 'The Past cannot change' comes to. What then about true statements concerning the Future? Well, if we can falsify a statement about the future, it simply is not true. If 'changing the Future' means only falsifying predictions, then we can surely change the Future. Why then can we not falsify retrodictions? The answer is we could, in one sense, do that. If I knew that someone would retrodict that I ate a peach at t-1, I could eat an apple instead, and so falsify this retrodiction. But this is just what I do not know. If we knew what the historians of the future would say about us, we could falsify their sentences if we wanted to, just as we can, if we wish, falsify what people before us have predicted that we shall do. Why do we not know the future in this sense? I am not able to say. But does that sentence of Peirce's, with which we began, mean any more than that we do not know what the historians of the future will say? 'The Future is open' says only that nobody has written the history of the Present.

IX

FUTURE—AND PAST—CONTINGENCIES

My final disposition of the Ideal Chronicler was philosophically shoddy. One can hardly suppose one has dealt adequately with the problem of historical foreknowledge—the supposed knowledge of the history of events before the events have happened—by first inventing the possibility of someone having such knowledge, and then arguing that there is no such knower on the grounds that one has only imagined there being one. The question is not whether there is in fact a being possessing such knowledge, but whether there can be such a being, and this turns on the question whether such knowledge is possible. After all, to take a comparable case, there are forms of scepticism which no one has ever held, but this fact alone does not constitute a refutation of these scepticisms. Moreover, I cannot even claim to have invented the possibility of a being equipped with historical foreknowledge. For knowledge of the future is sometimes credited to God insofar as omniscience is ascribed to Him. Speculative philosophers of history, indeed, have at times regarded the whole of history as exhibiting some divine plan which they regard as their duty to discern, or which they credit themselves with having already discerned in part, perhaps through revelation. So I must offer some rather more positive arguments if I wish seriously to reject the possibility of historical foreknowledge. These I shall seek to give in the present chapter.

One conclusion warranted, I believe, by my discussion of narrative sentences, is that frequently and almost typically, the actions of men are not intentional under those descriptions given of them by means of narrative sentences. This does not, of course, entail that reference to human purposes are historically unimportant. The actions of the Palatine Elector are intelligibly explained with reference to certain ambitions which were certainly entertained by him, namely to gain, and then, having once lost it, to regain the crown of Bohemia. His various negotiations with France and with England, his attempts to raise money and to secure alliances, may correctly be appreciated in the light of these ambitions, as constituting parts of what I have termed his *projects*. Yet his actions had, at every turn, consequences which he never intended and which, in view of our ignorance about the future, he *could* not have intended. Yet it is in view of these consequences, and in terms of their wider bearing upon the Thirty Years War, that his actions have acquired, in historical perspective, the significance they bear. We see them, briefly, in a way in which Frederick could not have, and certainly not at the time of their performance. He would, indeed, almost certainly have been horrified to learn the sorts of narrative sentences which were, in time, to cover his actions.

But this case is typical. It is a commonplace piece of poetic wisdom that we do not see ourselves as others do, that our image of ourselves is often signally different from the image of us held by others, that men constantly over- or under-estimate the quality of their accomplishments, their failures, and their dispositions. Such discrepancies are seldom decided in the individual's favour, for our criteria for assessing performances are by and large behaviouristic. These discrepancies are nowhere more marked than in history, where in the nature of the case we see a man's behaviour in the light of events future to his performances, and significant with respect to them. Historians have an advantage which the actor, and his own contemporaries, could not in principle have had. Historians have the unique privilege of seeing actions in temporal perspective. It is, accordingly, as I have repeatedly urged, a misguided lament to complain that we, being at a temporal remove from the actions which concern us as historians, cannot know them in the way in which a witness might have. For the whole point of history is *not* to know about actions as witnesses might, but as historians do, in connection with later events and as parts of temporal wholes. To wish away this singular advantage [1] would be silly, and historically disastrous, as well as unfulfill-able. It would, in analogy to Plato's image, be a wish to re-enter the cave where the future is still opaque. Men would give a great deal to be able to see their actions through the eyes of historians to come.

But since actions are covered by historians with descriptions which the actor himself could not have given of them at the time, and since the actions cannot be construed as intentional under *those* descriptions, it is somewhat difficult to know what to make of the free-will controversy in its historical applications. There is a natural temptation to suppose that 'is free' and 'is determined' are contradictory predictions, so that if one of them is false of a given action, the other must be true. But this thesis fits awkwardly with the fact that an action may be covered by any number of true descriptions, some of which are true narrative sentences, but only under some of these descriptions is the action intentional. It is, I think by common consent, a necessary condition for an action to be free that it be intentional. The analysis of intentions is complex and open to question, but that much, at least, may be granted. It may be granted, moreover, that some actions are intentional. Even a determinist may allow this, his claim being only that an action, even if intentional, is not for that reason alone free.[1] But let us suppose, however it is to be understood, that there are free acts, and that every such act is intentional. There remains the fact however, that these actions are only intentional under *some* descriptions, and that there are others under which they plainly are not intentional. If ' *a* is intentional' is a necessary condition for ' *a* is free', it follows that *a* is *not* free under those descriptions. But are we to say that it is therefore determined under these descriptions? It does not automatically follow save on the assumption that if an act is not free, it is determined. But the free-willist must now allow that the same action, admittedly free under some descriptions, is not free under others; and that even if 'not-free' is automatically to be understood as meaning 'determined', then the identical action is both free and determined. From which it would follow that these are sub-contrary predications which can accordingly both be true of the same action.

On the other hand, are we really prepared to accept the rule that 'not free' is synonymous with 'determined'? Commonly, one would think, a determinist has some independent analysis in mind of the predicate he wants to assert of every action. At the very least one would take him to mean something like this: every action is the effect of causes the presence of which is entailed by the fact that the action occurred, and that, given these causes, the action must occur. This minimal characterization, however, encounters some vexing problems when we think of actions as covered by narrative sentences.

Consider my earlier example in which Aristarchus is described as having anticipated Copernicus—a sentence one might find in any modern history of astronomy. Certainly one would not wish to say that anticipating Copernicus is something which Aristarchus *intended*. Hence *this* was not something which, by the free-willist's criterion, could be counted a free act. But could we say that *under this description* Aristarchus was *caused* to anticipate Copernicus; that all the causes for the action were present, and given that they were, the action *must* take place? We might say so. But would we not be committed then to say, at the same time, that Copernicus' later actions must take place as well? For it is certain that Aristarchus could not have truly anticipated Copernicus if Copernicus was not later to do the things which Aristarchus did before. But then, if Aristarchus' action is determined, that is, must happen given the causes that now hold, then (logically) Copernicus' action must also happen, it must be true *at that time* that Copernicus' action must happen. But is this to mean that all the causes of the later event were present at the same time as were the causes of the earlier event? This would be a bizarre consequence, entailing, amongst other things, that an historian who sought to find the causes of Copernicus' behaviour in the fourteenth century would be misdirected: to explain it, he would have to examine events which took place in the fourth century B.C.

The alternatives are not especially tempting. Suppose we wished to retain our faith that the causes of an action, some of them, at least, are to be discovered in the immediately antecedent temporal neighbourhood of the action itself, and that historians of science are not misguided when they seek to understand Copernicus' behaviour in the light of rather local causes. But if an event E-2 is (at least in part) determined by causes which occur *after* an earlier event E-1, to which it is narratively related, then not all the causes of E-1 *under the required narrative description* were present when E-1 occurred. For in particular the causes of the later event which is logically required by the narrative sentence are not present. This, I should think, would warp rather badly the predictive claims sometimes made by determinists. For supposing we have two temporally separated events, E-1 and E-2, related by a narrative sentence, and that some of the causes of E-2 occur after E-1. Then quite clearly, even an ideally complete inventory of the world up to the time of E-1 would not provide adequate information for infallibly predicting E-2, nor even E-1 under the narrative sentence. Hence there would be true descriptions of events under which they could not be predicted, even assuming perfect knowledge of that event's temporally antecedent causes. One *may* say that *E-1* was determined by causes which occurred after it occurred, but this reverses the causal relationship which has counter-intuitive results for history. It would entail that some of the causes of an event are to be sought for in periods after the event has occurred, for example, that a full explanation of Aristarchus' actions would require research into causal circumstances holding roughly at the time of Copernicus.

I think there are two alternatives available to the determinist, assuming he will accept the minimal characterization I have given of his position. The first is to regard narrative sentences as bastards, and contend that determinism can account for all non-narrative descriptions of events. This is an heroic move, but I am not so sure it is a tolerable one, for it is by means of narrative sentences, to begin with, that we express our temporal perceptions of the world. Should the determinists retort that there is something wrong to begin with in perceiving the world in a temporal way, we may simply point out that descriptions of events as causes of other events are, as I have shown, special cases of narrative sentences, and already belong to our temporal perception of the world. If the language of causation is illegitimate, what really is the determinist arguing about? The second alternative really comes to much the same thing. It consists simply in saying that there are descriptions of events under which those events are not determined. But the determinist is doubtless no more ready to identify 'is not determined' with 'is free' than the free-willist is to identify 'not-free' with 'is determined'. It is for these reasons that it is difficult to know what to make of the free-will controversy in its historical applications. As covered by narrative sentences, it is hard to say that actions are either determined or that they are free. Indeed, in the typical case, they seem to be neither.

But now there is a form of determinism which may dispense with the annoying questions of causality, known, sometimes, as 'logical determinism'. It rests its case upon certain allegedly timeless properties of sentences, namely their truth values. Assuming some version or other of the Correspondence Theory of Truth, the logical determinist makes the claim that (for example) should I read the *Times* at a certain time *t-1*, then the sentence 'Danto reads the *Times* at *t-1*' is true, and false if I do not. But then, if it is true at any time, it is true at every time, that is to say, in particular it has always been true, just as it always will be true, that 'Danto reads the *Times* at *t-1*'. It little matters whether anyone has ever uttered this sentence, or whether it has even been written down, or whether anyone in fact knows it. For from the fact that it is true, and from the fact that if a sentence is true, it cannot be false, it follows that the sentence cannot be false. But if it cannot be false, then it cannot have been the case that I did *not* read the *Times* at *t-1*. Briefly, whatever happens must happen, and whatever fails to happen could not have happened. Here there is no mention of causes. Logical determinism has been criticized, for instance by Ryle, as a case where

We slide . . . into thinking of anterior truths as causes of the happenings about which they are true, where the mere matter of their relative dates saves us from thinking of happenings as the effects of those truths which are posterior to them.[1]

But the logical determinist does not, I think, make this mistake. First because causes are events in time, and propositions are not; a sentence being true is not something which happens, but rather, if a sentence is true, what it is true of must happen. The logical determinist pretends to no knowledge of causes. Doubtless, he will argue, events have causes, something caused me to read the *Times* at *t-1*. But then again it is time-lessly true that this would cause me to read the *Times* at *t-1*, and hence timelessly true that I should read the *Times* at *t-1*, and hence impossible that I should not.[2] In so far as Ryle himself speaks of anterior truths, he is entangled in the position he wants to criticize. To get the flavour of the determinist's argument, one must look to the writings of fatalists. The hero of Diderot's novel *Jacques le Fataliste* expresses the point in one of his interminable dialogues with his master. After chasing to bed a group of ruffians, he returns to the conversation:

M: Jacques, quel diable d'homme est tu! Tu crois donc . .
J: Je ne crois ni ne decrois.
M: S'ils avaient refusé de se coucher?
J: Cela était impossible.
M: Pourquoi?
J: Parce qu'ils ne l'ont pas fait.[3]

Let us try to reconstruct the sort of argument presupposed by this curious position. To begin with, there is the supposition that every proposition is either true or false. But since a proposition cannot be both true and false, it follows that if is true, it cannot be false. This, together with rules of synonomy permit us to argue as follows:

(1) *p* is necessarily either true or false.
(2) If *p* is true, then it is impossible that *p* is false.
(3) If it is impossible that *p* is false, then it is impossible that *p* is not true.
(4) If it is impossible that *p* is not true, it is necessary that *p* is true.

∴ (5) If *p* is true, then it is necessary that *p* is true.

By similar argumentation, we get

∴ (6) If *p* is false, then it is necessary that *p* is false.

And from (1), (5), and (6) we get

∴ (7) It is necessary that *p* is true or it is necessary that *p* is false.

Now truth is here supposed to be a relationship of a semantical sort between a sentence and an event, and though events occur in time, the relationship is apparently time-independent. Moreover, if a sentence is true, this relation must hold. The determinist would doubtless find it absurd to say that a sentence is necessarily true, is about some event, but the event might not happen. So if the sentence is necessarily true, the event must necessarily happen. So, finally, if the event happens, it necessarily happens, and it is impossible that it *not* happen. Such, in general, is the logico-philosophical baggage unwittingly carried by Diderot's fatalistic hero.

Narrative sentences hold no terrors, and raise no difficulties, for this kind of determinism. Indeed, its claim is that future events are determined, once and for all, to happen as they will. Though logical determinism does not require, as I pointed out, that the true sentences ever should be written down or known, Jacques the Fatalist liked to believe they were. 'Tout ce qui nous arrive de bien et de mal ici-bas,' he cheerfully informs his Master, 'est ecrit la-haut.'[1] This sounds very like the I.C., and very like historical foreknowledge, where the history of events before the events have happened is known by some omniprescient being. 'Savez-vous, monsieur, quelque moyen d'effacer cette écriture?' Jacques asks. I should like to take seriously the implicit challenge in this question, and examine the logic of historical foreknowledge now in connection with logical determinism.

One version of logical determinism was attacked by Aristotle in an exceedingly complex passage in *De Interpretatione*. He felt that the determinist's argument, if sound, ruled out the possibility of action and the efficacy of human deliberation. There would, he seems to have felt, be no point in deliberation if everything necessarily happens the way it does, and he felt that there is a point in deliberating, and in following the action deemed best in the light of one's information.[1] Accordingly, he undertook to destroy the determinist's argument. And the way in which he is most commonly interpreted to have tried to destroy it is this: he insisted that (1) is false. His positive claim was that it is not necessarily the case that every proposition is either true or false. This, if so, would certainly damage the determinist's argument, for though it might still be valid, formally, there would be nothing compelling in the conclusion if one of the premisses should prove to be false. (1) is false, Aristotle believed, because there are exceptions to it. Of course, Aristotle might either have destroyed his opponent's argument, or have deduced his own position, by attacking another premiss, namely (2). For (2) is equivalent to its own contrapositive[2]

(2a) If it possible that *p* is false, then *p* is not true.

And (2a) is plainly ambiguous. It can be interpreted in (at least) two ways:

(2b) If it is possible that *p* is false, then *p* is false.
(2c) If it is possible that *p* is false, then *p* is not definitely true.

But (2b) is false so that its equivalent contrapositive (2a) is false, and since (2) is required by the determinist's argument, the argument is destroyed by this interpretation. But the interpretation (2c) is very nearly Aristotle's own position. He might reason if it is only possible that *p* is false, then it is also possible that it is true. But, of course, not *definitely* true. Nor, again, *definitely* false. So if there are propositions of which we might say that it is possible they are true and possible they are false, there are propositions which are not definitely true and not definitely false. This contradicts (1).

This, however, would almost certainly be challenged by the determinist, mainly by asking what interpretation is to be given to 'possible'. If Aristotle understands 'It is possible that p is true and it is possible that p is false' to mean only 'We are not sure whether p is true or whether it is false', then Aristotle is in turn saddled with a falsehood, that is, 'If we are not sure whether p is true or false, then p is not definitely either true or false.' This is, the determinist would claim, either a falsehood or a redundancy, the consequent only repeating, in a different way, what is contained in the antecedent. But if, on the other hand, Aristotle does not have an epistemological interpretation in mind—and the determinist has no quarrel with this at any rate—but means that p is really, objectively, neither true or false, then the determinist just rejects this interpretation. He will insist that for every value of p, 'p is possible' is false—on this interpretation. Hence (appealing to material implication) (2b) is true. So there is no difficulty in this rendering of (2a). On the other hand, the rendering of (2a) which is favourable to Aristotle's claim in fact pre-supposes the point Aristotle would want to argue to. Accordingly, the issue must be joined as before, with the determinist affirming, and Aristotle denying (1). And the only point in this logical digression is to make it clear that Aristotle, in so far as he is understood to be maintaining 'Some propositions are neither definitely true nor definitely false' is doing so in the deep and not in the mere epistemological sense—'in the nature of things, and not merely in relation to our knowledge or ignorance of things' as Richard Taylor phrases it.[1]

'In the nature of things . . .'? The determinist finds this puzzling. Does Aristotle mean to say that there are, in the world, situations which are neither one way nor the other? Situations s in which 's is F' and 's is not F' are neither of them true and neither of them false? Aristotle does *not* say this. He does not say that there *are* or that there ever *were* such situations. He writes:

In the case of that which is or which has taken place, propositions, whether positive or negative, must be true or false.[2]

This statement already shows that Aristotle has in mind no claim which is merely 'in relation to our knowledge or ignorance', for there are innumerably many propositions concerning past or present situations of which we cannot say whether they are true of false. Yet Aristotle says, plainly, that they *must* be one or the other. By contrast with propositions about what is or has been, Aristotle writes:

When the subject is individual and that which is predicated refers to the future, the case is altered (i.e. not the same—οὐχ ὁμοίως).[1]

And it is statements about the future, or those statements about the future which are about individuals, which constitute, for Aristotle, the sole exceptions to (1). But Aristotle is not saying that there will be situations, in time to come, which will be neither the one way or the other; that the present, and then the past, will someday contain ambiguous situations. For then *every* past and present situation would be ambiguous. But he has said that whatever is present or past is unambiguous in the required sense, namely, that sentences about these are definitely either true or false.

This is a puzzling teaching, for suppose we have a sentence '*s* will be *F*' which satisfies Aristotle's condition at 18a 35–36. Can we not say '"*s* will be *F*" will be true or false?' Aristotle might plead an ambiguity here. He might say that it depends upon whether a genuinely temporal use of 'will' is being made. If not, then the sentence says no more than what he himself says at 18a 27–28, and is true by definition. But if it makes a genuinely temporal use of 'will' then it is neither true nor false. Moreover, if it does make a genuinely temporal use of 'will' it is not equivalent to '"*s* will be *F*" is true or false'. For the former is about the future, and neither true nor false, while the latter—making a temporal use of 'is'—is about the present, and false. Aristotle could have said, alternatively, that the former sentence does entail the latter, that indeed, the latter has just the same meaning as the former, that both make genuine references to the future and *neither* of them is true or false. Now Aristotle might have made much the same point, I think, by simply regarding all propositions about the future as false. We are used, these days, to philosophical conflicts in which one group will hold, of a given class of sentences, that the members of this class are neither true nor false, and the other group will hold that these same members are false. The classical case of this is singular referring statements, like for instance, 'The present King of France is bald' when there is nothing for these to refer to. Aristotle may have found this other possibility undesirable, perhaps because he regarded the disjunction of a sentence together with its denial as true.[1] 'There will be a sea battle tomorrow or there won't be one' he held to be true, and indeed logically true. He might have felt that if both disjuncts are false, this would fit such a claim ill. But it is not easy to see that the view that both disjuncts are neither of them true or false fits any better. Meanwhile, had he regarded both a given future-sentence *and* its denial as equally false, and still wanted to save the logical truth of *p* or not-*p*, he might have undertaken a refined analysis of future-sentences calculated to exhibit that, as these occur in ordinary discourse, they can both be false—an analysis of the sort Russell gave in his celebrated Theory of Descriptions showing that ' The so-and-so is such-and-such' and its apparent denial 'The so-and-so is *not* such-and-such' are not really contradictories, one of which *must* be true. But I myself have no such analysis to offer.

It is tempting to ascribe to Aristotle the view that the existence of a designatum for any given use of a singular referring expression ('When the subject is individual . . .') is a necessary condition (a 'presupposition') for any sentence employing such an expression to admit of a definite truth value. Then '*s* will be *F*' and '*s* will not be *F*' will neither of them have a definite truth value since *s* does not exist, *is* not. I do not believe, however, that he would accept this. For then '*s* was *F*' and '*s* was not *F*' would, by the same criterion, be neither of them true or false, and this conflicts with 18a 27–28. Nor could we amplify the necessary condition as follows: the present or *past* existence of a designatum for a singular referring expression. For suppose we speak of the beauty of Napoleon's fifty-seventh wife. There was no such woman. Yet a statement about her would be a statement about the past and must be definitely true or false, if Aristotle is right. A sinologist who laboured to establish that there was never such a person as Lao Tze would hardly take it lightly were we to tell him his sentence is not true—even if we consoled him by saying that it is not false either. So the amplification does not help. But the necessary condition, as I stated it at first, rules out any statement about the past as having a definite truth value, while to insist, in the face of this, that sentences about the past *do* have definite truth values raises the question: why not then sentences about the future? So I do not think Aristotle could accept either the necessary condition or its amplification as a presupposition for singular propositions being true or false.

What I want to suggest is that Aristotle is taking time seriously, and if we appreciate his teaching in this light, it not merely stops being puzzling, but turns out to be just our ordinary way of looking at things. I shall try to reconstruct him along these lines, and will show that he is committed to something like the necessary conditions just outlined; a fact which explains our temptation to ascribe to him that view.

Let us speak of genuinely temporal uses of the words 'will be', 'is' and 'was'. We shall need six sentences now in order to achieve our analysis:

(1)	*s* is *F*,	(4)	*s* was not *F*,
(2)	*s* is not *F*,	(5)	*s* will be *F*,
(3)	*s* was *F*,	(6)	*s* will not be *F*.

Let us now specify when it is appropriate to use (5) and (6). I will say it is appropriate to use these when (1)–(4) are all of them false. And these are false in the special case where *s* has not taken place and is not now taking place. I shall say that when *s* has not taken place and is not taking place, (1)–(4) are all *time-false*. I shall say that it is appropriate to use (5) and (6) when (1)–(4) are all time-false. We may even say that when (1)–(4) are all time-false, (5) and (6) are *time-true*. Meanwhile, should any of (1)–(4) be time-true—on the grounds that *s* has taken place or is taking place, then (5) and (6) are time-false. I shall also say that if a sentence is time-false, it *is* false. But it does not follow that if it is time-true, it *is* true. In particular, say, (3) and (4) may be both of them time-true,[1] but one of them must be false.[2] Aristotle moreover may be understood as saying that when (5) and (6) are time-true, they are in no further sense true and in no further sense false. Briefly, (5) and (6) are both either time-false, and hence false, or they are both time-true, and neither true nor false.

But let us take the case where they are both time-true. There are three cases to consider:

(A)	*s* never takes place,
(B)	*s* takes place and is *F*,
(C)	*s* takes place and is not *F*.

Assuming (A), (5) and (6) are always time-true, and (1)–(4) are always time-false and hence *false*. But assuming (*B*) or (*C*) is the case (and they cannot both be the case), we have the following situation. At a given time (1) and (2) are time-true, and one of them is true and the other is false. At that time, (3)–(6) are time-false, and (5) and (6) are forever afterwards time-false and false. *After* that time, (3) and (4) are both time-true, and one of them is true and the other false. At *this* time, (1), (2), (5), (6) are all time-false, and hence false, and go on being false for ever. But, again, when (1)–(2) and (5)–(6) are all time-false, (3) and (4) go on being time-true for ever. Moreover, whichever of them is true goes on being true for ever, and whichever of them is false goes on being false for ever.

According to this analysis, there are no anterior truths except that, when (1)–(4) are time-false, (5) and (6) are, by definition, time-true, but not *otherwise* true. As I have pointed out, future-singulars are really never true, but they may be false, as indeed they always are when they are time-false. One *might* speak of them as becoming true, but this, strictly speaking, means only that at some time (the time at which they 'become true') (1) and (2) are time-true and one of them *is* true. But whether (1) and (2) ever become time-true (and one of them true) depends very much upon the way the world goes, and this in part depends, though perhaps not to the same extent as Aristotle supposed, upon human deliberations. We may speak of *making* certain propositions true or false, and accordingly say that the truth or falsity of propositions depends upon how things are, rather than, as the logical determinist would have it, the other way about.

This *is*, then, very much our common way of viewing truth, time, and the world. But Aristotle was always one to take seriously the common view of things. The trouble with the determinist's position is that, unlike Aristotle, he does not take time seriously. The logical determinist has an answer to Aristotle, and I shall take this up in a moment. But before doing so, I want to comment upon some of the implications of Aristotle's position, as I have attempted to reconstruct it for the problems which motivate this chapter.

One problem, which Aristotle's teaching raises, is that of how sentences in mixed tenses are to be evaluated. I am thinking, of course, of narrative sentences, sentences about the past, to be sure, but sentences whose truth or falsity is contingent upon the truth or falsity of some sentence about the future. If sentences about the future are neither true nor false—though nonetheless time-true—are we to say that these sentences, logically dependent upon them, are neither true nor false as well? Or are we to say that they are false because time-false—even though the event referred to by them has in fact already happened, so that the sentences in question ought, by rights, to be considered time-true and either the statement or its denial just *be* true?

The question has some importance for us. For if there genuinely are future contingencies, then, it seems to me, there are past contingencies as well: incompatible descriptions, so to speak, hovering over a given past event, unable to establish definite semantical relations with it until something happens in the future. Talleyrand begat a child who was, it turned out, to paint some celebrated pictures, including the *Mort de Sardanapale*. We may *now* say:

(1) Talleyrand begat the man who painted the *Mort de Sardanapale*.

Indeed, some such sentence may have been whispered about the galleries when that painting first was shown. But until the painting was made, are we to say that (1) was true or false?

Talleyrand had multiple offspring, but only his bastard son, Delacroix, painted our picture. With this information, let us parse (1) as:

(2) Talleyrand begat Delacroix and Delacroix painted the *Mort de Sardanapale*.

This is a little story, as we would expect, (1) being a narrative sentence. But formally, it is a conjunctive proposition, and, if true, entails:

(3) Delacroix painted the *Mort de Sardanapale*,

which is true if (2) is true. But suppose that (2) is asserted after the conception of the painter but before the execution of the painting. The only way of rendering (2) so as to make it time-true is:

(4) Talleyrand begat Delacroix and Delacroix will paint the *Mort de Sardana-pale*.

The question is whether (4) is true when it is time-true. It is hard to say that it is true, for if it were then:

(5) Delacroix will paint the *Mort de Sardanapale*.

But (5) is exactly the sort of sentence regarded by Aristotle as neither true nor false. So we must regard (4) as either false, or as neither true nor false. I do not think it makes much difference *which* we say. Either way, however, it is clear that (4) is not equivalent to (2). Since both (2) and (4) are renderings of (1) at different times, it follows that (1) is true at one time, and either false, or neither true nor false at another time.

Personally, it seems to me that if we are to allow simple propositions to have no definite truth value, there is no good reason to disallow indefinite truth values to compound propositions as well. Indeed, since negative propositions are regarded as compound, the negate of (5), since it is a sentence about the future, will, and indeed must, in Aristotle's view, be neither true nor false if (5) is neither true nor false. At any rate, it will turn out that any compound proposition which is neither true nor false must contain, as one of its propositional parts, a simple sentence which is neither true nor false, and hence, in Aristotle's analysis, a time-true future-referring singular proposition. I shall say that when any such compound proposition *also* contains a time-true, past-referring, singular proposition, the entire compound proposition, such as (4), expresses a *past contingency*. So not every time-true sentence about the past is true or false. If the future is open, the past cannot be utterly closed.

Now we may turn to the question of historical foreknowledge. The medievals were particularly concerned with this question, largely because of two tenets of Christian faith which they found it difficult to reconcile. The first was that God is omniscient, and the second was that man is free.[1] The assumption is that God knows everything. Hence he knows everything about the future. And, since *p* is entailed by '*a* knows *p*' it follows that, if God knows the future, whatever he knows is true, and if true, it cannot be false. Hence, if I read the *Times* this morning, it is true that 'Danto read the *Times* this morning' and that was known by God all along. But what choice then did I have? I could not but have read the *Times* this morning. To say I could have done differently is to deny omniscience to God or to reject the analysis of '*a* knows that *p*' in accordance with which that sentence entails *p*. Aristotle's teaching can salve the sense of dilemma raised by the problem just sketched. For if '*a* knows that *p*' does in fact entail *p* (and this is a crucial logical fact concerning knowledge), then the former sentence expresses a sufficient condition for the latter one. However, by parity of analysis, the latter expresses a necessary condition for the former. In particular, if p is not true, then 'a knows that p' cannot be true. But sentences regarding the future, in Aristotle's analysis, are not true: they are not true and they are not false. Hence, if p is a time-true future-referring singular proposition, God cannot know that p. Neither God nor anyone can know the future. Much the same result could be obtained if we regarded these same sentences all as false. For one cannot know what is not the case. God may still be then omniscient, if one wishes to maintain that thesis. For to say that some being is omniscient, as Richard Taylor has pointed out, is to say that it knows everything that *can* be known.[1] But then the future cannot be known. We can have rational beliefs, up to a point, about the future, but not knowledge. In this regard, God has no special advantage over the rest of us. Aristotle's teaching then rules out historical foreknowledge— though not necessarily historical fore-belief.

But if the future cannot be known, the past cannot be known completely. In so far as we are logically prohibited from having knowledge of the future, we are, again logically, prohibited from knowing whatever is expressed in sentences which express *past contingencies*. We can only know narrative sentences when these are time-true and true, and they are not this when they contain a singular time-true future-referring proposition. But since it is by means of narrative sentences that we ascribe historical significance to events, God, even if omniscient, cannot *know* what the significance of events is before they in fact *have* this significance. So in this regard again God has no special advantage over the rest of us. But in what sense then can history be said to conform to a divine plan, originally laid down? If Aristotle is right, as I have construed him, we can, I think, now furnish an answer to Diderot's fatalist. It may very well be that everything that happens is written down *la-haut*. This would be an interesting fact about the universe, but there would be no need to erase it, for not even God could know if it was right. Indeed, he could know rather quickly that it was false unless it changed tense with the happenings in the world. For it would very quickly become time-false, and hence otherwise false, if it were written in language which takes time seriously. But suppose, one might suggest, it were written in an untensed idiom? With this suggestion, we may now come to the final encounter between Aristotle and the logical determinist.

The logical determinist may feel himself to be justifiably puzzled by Aristotle's teaching that propositions referring to the future are neither true nor false, and even more so by my amplification of this teaching in terms of time-truth and time-falsehood. His claim is that a proposition, if true, is timelessly true, and that in general propositions are true (or false) independently of the time at which they may be asserted. But this is plainly false. There are indeed propositions for which this is the case, but these are propositions which do not require temporal factors amongst their truth conditions. But Aristotle is pointedly not speaking of such propositions: he speaks of those with respect to which time must be taken seriously. If I say that Smith left his house, and Smith *has not* left his house, then I have spoken falsely—even though *an instant afterwards* Smith leaves his house. We say that Smith's wife, answering the phone, and replying to an impatient query that Smith has left the house, knowing that Smith has not, but means to and is on the verge of doing so, has told a 'white' lie. But a lie, whatever its colour, is false. *Tensed* sentences do very much depend upon the time of their utterance in order to be true or false: the time of their utterance indeed can make the difference between truth and falsehood.

Now there is a standard counter to this claim. It consists in saying that tensed-sentences are incomplete as they stand, are really not sentences at all but sentence-functions with an implicit temporal variable. By making the variable explicit as in: 'Smith leaves the house at t-x' we get something which is not merely not true, but not false either, any more, say, than '$x + 3 = 9$' is true or false without giving a value to x. But now if we but put a constant for the variable as in: 'Smith leaves the house at t-1', we get a sentence which is either true or false, and, moreover, true or false independently of the time of assertion. In so far as a future-referring sentence does not specify the *time* of the event it refers to, it admittedly is neither true nor false: it just is not a sentence. And so for the grammatical alternatives in different tenses. Aristotle was unduly restrictive. But a sentence, even if grammatically in the future tense, when the time involved is made explicit, is then timelessly true or false. It is on sentences such as these, the logical determinist goes on to say, that he rests *his* claim.

But this counter-move has an obvious logical antidote. It is simply that the untensed sentence, 'Smith (tenselessly) leaves the house at t-1' does not tell us, as the tensed sentence 'Smith left the house' does, that the action has already taken place, is in the past. To be sure, in a given context, in which speaker and hearer know the time referred to in the sentence, as well as the time at which the sentence is asserted, the hearer may, by doing some calculation, conclude that the action took place in the past. But this remains an extra piece of information which the need for specifying contexts already shows has not been quite incorporated into the sentence itself. In so far as this is so, the determinist cannot legitimately claim to have eliminated tensed idiom: he had only, so to speak, relocated it in the context; this is what we might expect in view of the conclusions of Chapter IV. We cannot give timeless equivalents for sentences which include temporal circumstances amongst their truth-conditions. So this move fails.

Aristotle may press the point further, for does the determinist's argument not require some version of the Correspondence Theory of Truth? If it does, then what analysis is he to give of the claim that a sentence in the future tense *is* true? The Correspondence Theory requires that there is something, say an event, which corresponds, however this is understood, with the sentence in question. But *is* there anything which corresponds with a future tensed sentence? If there *is*, in what sense are we properly using the future tense? If the determinist answers that there *is* not, but *will be* a factual correspondent to the sentence, is this not employing tensed language? To be sure, the determinist might argue that a future-tensed sentence may be understood to mean that a given event is timelessly later than a given reference event; but then we revert to the first argument, for the Thirty Years War is timelessly later than the Defeat of the Spanish Armada. The question, however, would always remain how we should enable someone to know that the latter event is future without telling him whether the former event has occurred or not. We can only say that the latter event is *future to* the former event, and that an event may be *future to* another event without being future, for in fact the Thirty Years War, though future to the defeat of the Spanish Armada, is *past*.

So whichever way we look at the matter, the determinist either fails altogether to find tenseless equivalents for tensed propositions, or else, in thinking he has succeeded in doing so, implicitly smuggles in the precise sort of temporal information he meant to eliminate. Moreover, the very statement of his own position requires the sort of temporal information of which he boasted that his position is independent. He cannot get clear of time. Even if he should point out to Aristotle that *his* (or my) specification of the truth-conditions for tensed sentences pre-supposes temporal information—a past-tensed sentence is time-true if the event referred to *has* happened—this need not worry Aristotle. It strengthens, if anything, what I take to be his implicit claim, namely that we cannot speak of time without using temporal notions to begin with. There are perhaps problems enough about time, but these must be faced jointly by Aristotle and his opponent in *De Interpretatione*. One *may* find it puzzling that there should be past, present, and future; that there should be asymmetries in time reflected as asymmetries in the truth-conditions for sentences which take time seriously, but it would always be a good idea, when such puzzlement is felt, to specify what it would be like for time to have what is puzzling in it be removed. In the end it may turn out that a statement of our puzzle about time is really simply a description of the way time is. The logical determinist's timeless notion of truth turns upon a timeless notion of *time*. So what can he mean in saying that the *future* is determined?

One may go on being a determinist in spite of my previous statements. One may quarrel with Aristotle that deliberation makes a difference, but the issue must be joined elsewhere, I think. Aristotle, I submit, is right in the controversy just discussed, and his being right entails the impos-sibility of historical *foreknowledge*; but this does not mean that we cannot predict events within limits. To this matter I shall turn in the next two chapters, for it is a question which can best be discussed in the context of the problems of historical explanation.

X

HISTORICAL EXPLANATION:
THE PROBLEM OF GENERAL LAWS

Towards the end of ch. VII, in considering the possibility that one way of distinguishing between 'plain' and 'significant' narratives might be that the latter, but not the former, offer explanations of what happened, I contended that a narrative already is, in the nature of the case, a *form* of explanation. In this chapter and the next I propose to defend this view.

To begin with, it seems a plain fact that in a good many cases, and unless he is committed to some general historical theory such as Marxism, an historian would spontaneously offer us a narrative when we ask him to explain for us a certain occurrence; and that he himself, when he wishes to find an explanation of an occurrence, will undertake to ascertain what we sometimes call the 'story'—meaning, roughly, the events which lead up to the event in question. But this is not uniquely the case with historians. If someone, for instance, has an automobile accident, and is asked afterwards to explain why (or to explain how) it happened, the answer naturally expected would be a narrative. Thus there is some justified inclination to say that historical explanations are simply narratives, and insist that this is all that 'explanation' is taken to mean in historical contexts. Since a narrative seems not merely to explain an occurrence, but to tell what happened over a stretch of time, there is again an inclination to say that telling what happened and explaining what happened are jobs which are done simultaneously, that in so far as a narrative explains, it also tells precisely what happened, and that in so far as it tells precisely what happened, it also, and at the same time, explains. In this way narrative description and historical explanation are of a piece.

Yet it may be legitimately objected that there is no contradiction in the complaint that one knows what happened, though one cannot *explain* what happened. Generally, this will mean only that one knows that a terminal event in a sequence of events has taken place, but that one is in ignorance of the precise sequence of events it terminates. A police officer, for example, were one to reply to his question 'What happened?' by saying that 'An accident took place', might correctly say that he *knows* this, but that this is not the 'What happened' he is interested in. He wants to know the story of the accident; he wants, at once, to know what happened and why it happened. The following account would normally be taken to satisfy both demands (so far as it is correct here to speak of *two* demands, for my claim is that these are only different ways of making the *same* demand):

The car was driving East behind a truck; the truck veered left; the driver of the car thought the truck was making a left turn, and proceeded to pass on the right; but the truck then sharply veered to the right, for it had gone left to make a difficult right turn into an intersection which the driver of car had not seen; and so there was a collision.

Were the officer to persist, after this, in saying that he knows what happened, but still wants an explanation, we should, I think, be puzzled. We could have given a more detailed account, but hardly a clearer one; and it is difficult to think of any further detail which would make it any clearer why the accident took place. What more could he want? To be sure, we abstracted from the total set of happenings which occupied a stretch of time terminating with the collision a very select set, but this was because our criteria of relevance were operative, and certain events, though part of that stretch of time, are not to be counted part of the *story*, nor (to put it differently) part of the explanation. Perhaps at the moment the truck veered left the radio in the car was playing the Appassionata Sonata. This is not mentioned for it lacks significance. It might, of course, have had significance. The driver, a musician, was so intent upon a competitor's performance of a piece in his own repertoire that his attention was distracted from the conditions of the road. But this was not the case, and is at any rate ruled out by the claim that the driver thought: the truck is turning left. To be sure, this may have been a lie, and he may have been absorbed in the music, but then, in lying, he would not merely have failed to give the correct explanation, he would also have failed to say what happened. There is, therefore, some *prima facie* justification for saying that to tell what happened, by means of a narrative description, and to explain why something happened, is to do one and the same thing, and that a correct explanation of *E* is simply a true story with *E* as a final episode.

It may be argued, however, on the basis of considerations advanced in our discussion of narrative sentences, that there may be any number of true stories with E as a final episode, and that not all of these true stories would necessarily be counted, in a given context, as an explanation of E. Again, it might be argued, that there might be explanations of E which do not naturally fall into the form of a story. Indeed, the position has seriously been defended that a correct explanation has quite a different form, the form, namely, of a deductive argument, with a sentence describing E as its conclusion. So the sociological fact that narratives are often, and even typically, advanced as explanations, and the psychological fact that often, and even typically, we want, and are only satisfied with, a true story when we require an explanation, cannot be accepted as fully supporting, without further analysis, the claim I seek to defend. Before defending it, however, I should like to examine some rival analyses of the concept of historical explanation; for the fact is that few problems in the philosophy of history have received the sort of concentrated philosophical scrutiny that this one has, and by and large the discussion has been concerned with the adequacy of an analysis of historical explanation radically different from the one I am committed to defend. Accordingly, I shall commence with an analysis of that controversy which I hope to be able to resolve. Then I shall return to the analysis of stories.

The mooted candidate for an acceptable analysis of historical explanation, and for explanation in general, has been advanced, in its classical form, by Professor C. G. Hempel.[1] The main controversial element in his analysis centres about his insistence that, amongst other necessary conditions for an explanation e being an *adequate* explanation is this: e must include at least one general law. We may map the terrain of philosophical conflict which this thesis of Hempel's has given rise to by taking it in conjunction with two purported facts having to do with historical practice. The problem of historical explanation, as it is currently debated, emerges as a result of logical tensions amongst the following three propositions:

(1) Historians sometimes explain events.
(2) Every explanation must include at least one general law.
(3) The explanations historians give do not include general laws.

The difficulty is that while most of us would be prepared, with perhaps some qualification, to assent to all three of these, we are logically prevented from assenting to more than *two*, at least as they have been stated here. For to no matter which pair of these we choose to assent, these will together entail the falsity of the remaining one. Specifically, (1) and (2) entail not-(3); (1) and (3) entail not-(2); and (2) and (3) entail not-(1). Hence, there are three possible positions to be taken, though we can in fact reduce this number by supposing it to be common ground that (3) is true. Moreover, (3) can be established as true, or true for the most part, by examining instances of purported explanations, as furnished by historians. It will then be seen that these almost never include general laws (it will be seen, in fact, these are almost always narratives). So the question then is which of (1) or (2) are we prepared to assert in conjunction with (3), given that we cannot, as they stand, assert *both*. Which of these choices we make will depend, in some measure, on our general philosophical commitments, and whether we take, as paradigmatic, the actual claims and practices of historians, or the claims and practices of logicians. Thus, to employ the mythical types often used as characters in the dramatic confrontations by means of which these questions are sometimes worked out in philosophical writings, the Logician will assert (2) and deny (1), and the Plain Man (in this case the Historian) will assert (1) and deny (2).[1] Here the division might be ultimate in the way in which philosophical differences so frequently appear, with arguments marshalled against counter-arguments in elegant batteries facing one another across a dialectical chasm, were it not that in some ways there might seem, to a moderate person, some truth in *both* positions. Eager as one might be to defend one of them, it would really be embarrassing to deny the other. So in effect, we can specify two radical, and two moderate positions, which might be taken, only assuming that (3) continues to be regarded as common ground:—

(A) (2) is absolutely true and (1) is absolutely false.

(B) (2) is absolutely true, and (1) can be restated in an acceptable way, though it is false as it stands.

(C) (1) is absolutely true, and (2) can be restated in an acceptable way, though it is false as it stands.

(D) (1) is absolutely true and (2) is absolutely false.

Obviously (*A*) and (*D*) are radical, and (*B*) and (*C*) moderate positions.

Moreover, two moderate philosophers may adopt the same position, in the terms in which I have stated them, and yet go on to disagree violently with one another, for there is considerable room for difference over the manner in which the required restatement is to be made. Even amongst the radicals, there might be considerable difference in the *reasons* they would accept for ruling out (1) or (2) as absolutely false. So philosophical differences, as well as different ways of appreciating historical practice, re-appear even in this somewhat refined statement of the quarrel. I shall only try to identify, and then comment upon, one representative of each of the four positions. Each of these, incidentally, may be regarded as a different way of accounting for the purported truth of (3).

(A) It is not easy to identify a *single* philosopher who would defend (A), but it is by and large the view of those thinkers whom I shall identify as Historical Idealists, e.g. Croce, Dilthey, Collingwood, etc. These, whatever individual differences there may be amongst them, are unanimous in insisting upon a radical distinction between the behaviour of human beings and non-human entities, and a correspondingly radical distinction between the groups of disciplines which respectively study these two allegedly distinct kinds of behaviour. These are termed *human* sciences, in contrast with the *natural* sciences, or, to use the familiar German expressions, *Geisteswissenschaften* in contrast with *Naturwissenschaften*. It is, according to this school, the task of the natural sciences to explain non-human phenomena, and indeed to do so by means of identifying the laws which these phenomena invariably conform to. Hence to explain is to bring under a law, and (2), accordingly, is absolutely true. But human beings do not, characteristically and essentially, act in accordance with general laws. Human beings are free agents, historical events are unique and unduplicated, and the action of human agents must be appreciated in the light of certain inner occurrences, such as, purposes, motives, desires, which cannot be asserted of non-human entities.[1] It is the task of the *human* sciences to reconstruct these inner mechanisms which, since they cannot be observed must be reached in some other way. Only external behaviour can be *observed* and is unintelligible, in the case of humans, except by reference to unobservable occurrences in the agent's minds. This process consists in some sort of emphatic apprehension of the inner workings of another mind, a process often termed 'understanding', or, again to use the familiar German word, *Verstehen*. Hence (1) is absolutely false. Historians, as *human* scientists, do *not* explain—if 'explain' is understood in the sense claimed to be appropriate to the *Naturwissenschaften*. Rather, historians 'understand' the unique and never duplicated episodes in which free human agents have engaged down the ages. (3), then, is trivially true. Historians' explanations do not include laws for the simple reason that there are no historical explanations. The reason we do not find general laws mentioned in history books is because there are no explanations, in the strict *Naturwissenschaften* sense of the term to be found there. This is not a defect, however, and to demand that historians explain is logically to misunderstand the nature of their discipline, as well as metaphysically to misunderstand the nature of human beings and the important differences between them and other sorts of beings.

(B) It must in some measure have been in response to the sorts of views roughly sketched under (A) that Hempel wrote his celebrated paper on the function of general laws in history. For one of the views subscribed to by the *Logical Empiricist* school, of which Hempel was an outstandingly active and creative member, was the *unity of scientific method*; the view, namely, that differences amongst kinds of phenomenon need not be reflected as differences in the scientific representation of the properties and behaviour in the subject-matter, and that scientific method is invariant as to subject. Strictly speaking, Hempel was not concerned in a direct way with phenomena, but—consistent with the way in which the Logical Empiricists viewed their job—with the language scientists used to describe phenomena. Now differences in subject-matter will indeed be reflected in science as a difference in scientific *vocabulary*, or more precisely, as differences in the so-called 'non-logical' vocabulary of this science or that, but this in no way bears on the logical structure of sciences which employ different non-logical vocabularies. There was, it is true, a tendency to regard the basic vocabulary of any science as explicitly definable by means of *observational* terms, or as reducible to sentences only making use of observational vocabulary; a programme of analysis which would oppose the idealist's claim that human behaviour can only be understood in the light of intrinsically *un*observable events. However, the thesis of physicalism need not be presupposed by the thesis of the unity of scientific method, and one of the things Hempel meant to show was that explanation, in particular, has exactly the same structure whether it has to do with human or non-human behaviour. In addition, of course, he meant to show that we can, and *do*, explain human behaviour.

To *explain* some phenomenon, as Hempel saw it, is to perform some operation on a *sentence* (the *explanandum*) which describes the phenomenon in question. Hempel's view was that this operation consists in *deducing* this sentence from some set of premisses to be taken as adequate grounds for the explanandum (if we do not have adequate grounds, how can we be said to have explained?). The set of premisses he termed the *explanans*. I shall term this the *Deduction Assumption*, though it is plain that the logical feature of deducibility from premisses cannot be regarded as more than a necessary condition for explanations, even if we grant that an adequate explanation must be representable as a deduction. For it is surely always possible to find premisses from which the explanandum may be deduced, though no one would regard this as an *explanation* at all. Nevertheless, from the Deduction Assumption, together with our pre-analytical notion of explanatory *in*adequacy, we may, I think, elicit the remainder of Hempel's analysis by means of a sort of transcendental argument.

(i) The explanandum, since it describes a particular occurrence, is singular in form. Let it be the sentence Ga. Now the Deduction Assumption alone tells us very little about the formal composition of the premisses which are to make up the explanans for Ga, except that, what ever they are, they (by definition) cannot be consistently true while Ga is false. But so many propositions might satisfy this condition: Ga is a deductive consequence of itself, of its own double negation, of any conjunction in which it itself is a conjunct, etc. No one would regard these deductions as explanatory, however, and without ensnarling ourselves in logical niceties, let us suppose that we ordinarily explain an event with reference to its conditions, so that one of the premisses must certainly describe some condition for the event which Ga describes, and this premiss must then be distinct in form from Ga. Since the condition is a *specific* condition, it, too, demands a singular proposition to express it. Let this proposition be Fa.

(ii) But of course, *Fa* in no sense deductively entails *Ga*. So if the premisses consisted uniquely of *Fa*, the Deduction Assumption would be violated. But notice that *Fa* must express a condition sufficient for the occurrence described with *Ga*, for otherwise *Fa* might be true and *Ga* false, and we would then not have an adequate explanation of why *Ga* is true. If the event might not have occurred even though the condition held, the holding of the condition fails to explain why it *did* occur, and the explanation is thus inadequate. So we need to specify that the condition is a sufficient condition, and by definition, to say that *Fa* is a sufficient condition for *Ga* is to assert that $Fa \supset Ga$. But *this* conditional sentence, together with *Fa*, really does yield *Ga* as a deductive consequence.

(iii) Let us imagine a condition *Fb*, indiscernible from *Fa*, except that when *Fb* holds, *Gb* does not, though when *Fa* holds, so does *Ga*. In other words, supposing that *a* and *b* are alike, and that the condition *Fa* and *Fb* are alike, but when *Fb*, ' *Gb* ' is false, though when *Fa*, ' *Ga* ' is true. And, for the sake of simplicity, let us suppose that everything else is the same. Well, if we do suppose that, then either *Fa* and *Fb* are not indiscernible, or else *Fa* is not a sufficient condition for *Ga*. That is to say, unless, for every *x*, $Fx \supset Gx$—everything else being the same—then *Fa* is not a sufficient condition for *Ga*. To suppose that we do have a sufficient condition, then, is to commit oneself to the general proposition that under similar circumstances, the same things will happen when the same conditions hold. And indeed, if this general proposition is false, then we don't have a sufficient condition; and if we don't have that, we don't have an explanation. Plainly, then, our explanation really requires this general conditional sentence. Let this be $(x)(Fx \supset Gx)$. And since we can easily get $Fa \supset Ga$ from this by a well-known rule of inference, the latter is not independent of the former, and can be dropped as one of the premisses.

(iv) The explanans, then, minimally consists of (at least) two premisses, *Fa* and $(x)(Fx \supset Gx)$. But of course, the empirical interpretation of the latter is as a *general law*, and the empirical interpretation of the former is as an *initial condition*. Hence, to (empirically) explain an event is to connect that event with a condition, and by means of a law, which is what Hempel has represented. This satisfies both the Deduction Assumption and our pre-analytical notions.

Notice that what we have explicated here is the concept of an *adequate* explanation. But is an inadequate explanation really an explanation at all? If it is not, any more than a false face is a face, we need not qualify the term 'explanation' with 'adequate' any more than we need qualify the term 'face' with 'true'. And we will thus have explicated, by means of a transcendental argument, the concept of explanation.

Hempel and Oppenheim[1] specified a variety of further considerations, syntactical and semantical; but there is little point listing these here, and I shall merely indicate the manner in which their analysis was applied by Hempel to history.

To begin with, assuming he had given an explication of the concept of *scientific* explanation, it was plain that science could give explanations of single events. Given that no pair of events belong to all the same classes, it follows that every event is different in some degree from every other, while, from the fact that any pair of events share at least one property, it follows that no event is unique. In this regard there is no room for distinguishing history from the natural sciences on the ground that the former deals with single events, for so does science; nor on the grounds that it deals with unique events, for there are none. But it might be objected that such properties as a given historical event may share with other events are trivial or uninteresting, that there is a sense in which historical events are undeniably unique. How then might they be covered by general laws of the sort exacted by Hempel's analysis? Well, the fact is that historians themselves, when they undertake to explain events, while they do not precisely mention any laws, nonetheless tacitly presuppose their existence, their explanations accordingly being in the nature of elaborate enthymemes. On the other hand, and this is part of the relative backwardness of some of the social sciences, the fact remains that such laws as might be implicit and presupposed cannot be explicitly stated in wholly unexceptionable form. So, strictly speaking, if we mean by 'explanation', exhibiting an explanandum as a deductive consequence of an explanans containing scientifically acceptable general laws, then, of course, (1) is *false*. But we may restate (1). We may say that the explanations historians offer are really not explanations as such, but are, in Hempel's phrase, 'explanation sketches'. Places are, so to speak, marked off where the appropriate general law which is presupposed, will, in time, be inserted, converting the sketch into a fully satisfactory explanation. Such a sketch

consists of a more or less vague indication of the laws and initial conditions considered as relevant, and needs 'filling out' in order to turn into a full-fledged explanation.[1]

Along these lines, then, Hempel argues that (2) is true, (1) is true if we replace 'explain' with 'sketch explanations for'; and (3) is true if we realize that we are talking about *explanation sketches* which do not *include* general laws, but which presuppose them.

The general analysis here exhibited is apparently due, originally, to Professor Karl Popper,[2] and can be said to enjoy a wide acceptance amongst empiricist philosophers, as well as amongst certain avant-garde historians.

If I am correct in saying that Hempel's position was a direct rebuttal of (*A*), the remaining two general positions have typically been occupied by philosophers concerned to rebut (*B*) in Hempel's formulation. In this sense, Hempel's analysis has determined the complexion of the subsequent history of the problem. Parenthetically, it would be a neat problem to try to explain, along Hempelian lines, the subsequent state of philosophical discussion which his own analysis clearly determined. But let us now turn to the moderate, and then the radical criticism of (*B*).

(*C*) Certainly what typically is sought for in historical explanations are the causes of an event, and certainly to assert that something *K* caused an event *E* is to commit oneself to the existence of some general law to the effect that *K*-like events cause *E*-like events. So much is covered by the sense of 'relevance' identified as operative in my earlier example. In so far as we indicate causes, or supposed causes of an event, we are clearly doing something which conforms to Hempel's notion of an explanation sketch. But have we any right to suppose that the sketch in question requires only explicit citation of the law appealed to in order to qualify as a fully-fledged explanation? For the law may, as many have pointed out, be a probability law. We may know that *E* occurred, and that *K* occurred, and regard it as likely that *K* caused *E* on the grounds that frequently *K*- and *E*-like events are joined. But suppose there are known cases where an *E*-like event has not been preceded by a *K*-like event, though in the main the connection holds. From a sentence describing *K* we cannot, by appeal to such a probability-law, strictly *deduce* a sentence describing K. We do not, strictly speaking, have adequate grounds for the explanandum '. . . E. . .' for it is now logically possible for this sentence to be false and '. . . K. . .', as well as the probability law, to be true. In this sense we would then not have explained E even if we made the law explicit, and it may very well be the case that the only laws presupposed in historical explanations are of this kind, so that we cannot, by appeal to these, ever succeed in explaining events. Meanwhile, if there is some other law of strictly universal validity, the fact is that we do not know it and can hardly be said, in any obvious sense, to have presupposed it, and at any rate it could not be a law connecting K with E in as much as, by hypothesis, they are connected only by a probability law. So a different initial condition would be required from the one we refer to, and accordingly ours cannot be regarded any longer as a sketch for the final explanation, needing only to be 'filled in'. Rather, it would have to be scrapped. On the other hand, it may simply be that probability laws are ultimate, in which case nothing better is to be hoped for, and the filling in still does not yield an explanation in the stipulated Hempelian sense. But this may be the case not merely in history and in ordinary life, but in science as well, and we should then be obliged to conclude that perhaps neither plain men nor historians nor scientists have *ever* succeeded in explaining any phenomena at all. It now becomes a question as to whether we have failed so abysmally in our explanatory activities, or whether the criteria for an explanation have become, in Hempel's analysis, so exalted as never to be satisfied. Perhaps the deducibility assumption ought to be abandoned. But then, so far as we abandon it, we abandon, as well, the grounds for Hempel's entire analysis, which very nearly follows, as I outlined, as a logical consequence of this assumption. A closer look might be taken as well at the *use* of the word 'explain'. Surely, in one sense, and perhaps in the main sense of this term, historians, plain men, and scientists succeed in explaining things. They 'make clear', they produce 'understanding', of the things and events which before were dark and not understood. But then in this sense (1) is unquestionably true. Historians furnish understanding, they make clear to us why things came about as they did. Hempel, fastidious but misdirected in his semantical and syntactical stipulations, quite overlooked the central *pragmatic* dimension of the notion of explaining.

Yet (2) is not *absolutely* false. It is false in so far as it suggests that the laws in question are categorical, and enter into explanations as major premises in deductive processes. For the laws might, as we have seen, be only probability assertions which could not at any rate make the required deductive connections. But suppose we continue to question the Deduction Assumption: we might still give an acceptable sense to the idea that laws are included or involved in explanations, and so restate (2) as to make it unexceptionable, and compatible, at the same time, with the pragmatic aspect of explanation. For example, when a man offers to explain E with reference to K, he may be called upon to *justify* the explanation. Then indeed he might have to cite some general law, or to indicate roughly what law it is in virtue of which the explanation is apposite. But is there any reason to regard such a law as *part* of the explanation? The answer is no. The law may indeed be part of the *grounds* for the explanation, and failure to produce a law may expose one to the charge of having advanced a groundless explanation. But then there are many distinct kinds of grounds for any explanation; there being no need to include all of them, or to exclude any if some be included. For example, when we cite K in explanation of E, certainly we must have some confidence that '... K ...' is true, but are we to include our evidence for this belief as part of the explanation? Even Hempel did not go so far. Why then go so far as he went in demanding that any grounds at all must be reckoned part of the explanation?

So if we understand 'includes a general law' as meaning only that the explanation includes, amongst its *grounds*, at least one acceptable general law, then (2) is not merely philosophically, unexceptionable it is plainly *true*: Or nearly so. For what we might at best hope to offer, in the request for this sort of backing for an explanation, is some general law-like sentence which may not quite be a law in the required Hempelian sense, unless, indeed, we loosen somewhat our criteria for laws, and consider general statements to be laws even when they allow of exceptions and extenuations. Thus we may seek an explanation for Jones building a fire. We may explain it with reference to his feeling cold. Can I provide an exception-tight law to justify this? Not easily. People feel cold without building fires (ascetics) or build fire without feeling cold (janitors), and people both feel cold and build fires without doing the latter *because* of the former (it is their job, but they have no right to build fires on their own behalf). Yet there is no question that we accept as a truism that people build fires when they feel cold. We appeal to a 'law' which might run as follows: people (generally) when they feel (sufficiently) cold (tend on the whole to) build fires (of one sort or another) (more or less). It is such general truistic sentences which historians might properly be said to imply, and with which they might justify their explanations. But we cannot, by tinkering, remove the parenthetical qualifications in these and transform them into laws of nature.

This position, and some of the arguments in its favour, has been seriously advanced by Michael Scriven.[1] Professor Karl Popper was the first to draw special attention to the historical employment of trivial generalizations,[2] but, unlike Scriven, considered their rôle deductive rather than justificatory. Ernest Nagel has emphasized the rôle of probabilistic laws, and while aware they do not permit deductions, none the less considers them, unlike Scriven, as part of the explanation and not merely part of the *grounds* of explanation.[3] That so-called laws of nature, even in science (where we would most naturally seek for them) are highly idealized, and do not precisely hold in any given context where allowances have to be made of a *caeteris paribus* nature is a commonplace in the philosophy of *applied* science. That most of the generalizations which cover human behaviour are truisms may be confirmed by consulting nearly any sociological study chosen at random.

(D) It is not easy to identify a radical critic of Hempel, a philosopher who is unreservedly committed to the *absolute* falsity of (2), but perhaps Professor William Dray of all the main writers on this problem is closest to this view. Scriven who raises doubts as to whether there are ever laws of the required sort, nonetheless concedes that some sentences ('normic sentences') are, as ground for explanations, 'involved' in explanation. This might strike Dray as too much of a concession. He writes that 'in any ordinary sense of the word, the historian may use *no law at all*',[4] and his italics suggest that this is his radical view of the matter. There are more guarded moments in his book, however, where he suggests this is perhaps excessively rash. But for all that, the direction of his destructive argument seems to point toward the italicized phrase, and his avowed purpose of destroying what he terms the 'Covering Law Model' encourages me to locate him here.[1]

Whether or not 'in any ordinary sense of the word' historians use laws is, of course, a factual question, open to empirical investigation, and this, as Dray realizes is not the kind of question that is being raised. The question is, even supposing they did use laws, whether their use of laws is in any way a necessary condition for any explanation they might give, and whether their use of laws would constitute a sufficient condition for an explanation. These questions Dray answers negatively. His argument rests upon the presumed psychological fact that an historian may reasonably stick by an explanation he has given without feeling himself to be committed to the independent truth of any law which might be proposed as covering the event, so that in some loosely specified sense of entailment, no law is *entailed* by his explanation. But again, and for roughly comparable reasons, an historian might accept a given law as independently true, and even as covering an event in question, without regarding it as explaining the event it may in fact cover.[1] To illustrate the first case: explaining the fact that Louis XIV died unpopular by showing that he pursued policies detrimental to French national interests, the historian may find it difficult indeed to say what law it is which sanctions this explanation, and may challenge with impunity any logician, who argues that there *must* be a law, to state what the law is. The explanation does not entail a law in any obvious way, and for all that the historian is *certain* that he has correctly explained the fact. To illustrate the second case: concerned to explain the fact that Sir Brian Tuke was bow-legged, an historian would (Dray claims) find it unenlightening were a colleague to point out that all medieval knights were bow-legged.

If we reject the so-called Covering Law Model of historical explanation, what model shall we adopt? Dray suggests, with many reservations, what he speaks of as a Continuous Series Model:[2] we explain a gross event by sundering it into a series of sub-events until we have reached some set of sub-events which are just understood, events at which we 'doff our hats', and of which no explanation is wanted or needed. Yet Dray hesitates to sponsor this model with any marked enthusiasm, first because there are internal difficulties, but secondly because there is no reason to assume that there *is* any one model which every accepted instance of historical explanation conforms to. Historians may provide explanations of widely differing forms, and Dray's programme, to judge from the rest of his book as well as his subsequent publications,[1] has been to offer a phenomenology, so to speak, of historical explanations. Like Scriven, meanwhile, Dray is insistent upon the pragmatic aspects of explanations. Explanation is always relative to a context and to a level of knowledge already possessed.

If I am right in committing Dray to the absolutely falsity of (2) and the absolute truth of (1), together with his critical arguments, we can see what reason he would give for the truth of (3), quite independently of what evidence in support of it one might produce from a scrutiny of historical writings. It is simply that we ought never to have expected to find laws in historians' explanations to begin with. We can account for the lack in this fashion, but we must refer to the history of the problem, and to the misleadingness of the Covering Law model, to account for the fact that anyone thought this lack worthy even of remark.[2]

The above, then, are some of the ways in which the main positions on the problem of historical explanation have been articulated. Even with these brief accounts, it should be clear that the partisans in this strife have often tended to approach the problem within a context of differing philosophical commitments, and with differing criteria of what constitutes a solution to a philosophical problem. This permits a good deal of non-lethal cross-fire, and a good deal of ground-shifting, without it being wholly clear how the problem might be solved to everyone's satisfaction. In part, this is due to the fact that much of the difference between this philosopher and that is due to straight-forward verbal disagreement, the contestants really quarrelling over how a certain key word is to be used. Professor Scriven, for instance, has contended that an explanation is whatever furnishes *understanding* of the phenomenon or event which wants explaining. There is no reason why Professor Hempel need contest this claim. He might ask, however, for an analysis of 'understanding', and might go on to say that he would recognize only that as providing understanding (in contrast with pseudo-understanding?) which he has said all (genuine) explanations exhibit: a deductive structure. As regards Professor Dray's reservations on the explanatory value of 'All medieval knights were bowlegged' with regard to the fact that Sir Brian Tuke was bowlegged, Professor Hempel might say that there is a question to begin with whether this sentence satisfies the criteria for a general law, and that he is as aware as anyone of the difficulties in saying what these criteria are. Even so, the sentence, if true, is hardly so non-explanatory as Dray suggests. It indicates the direction in which we might seek for an explanation, for example, Sir Brian's position as a knight, in contrast, say, with a comparable general sentence 'All members of the Tuke family were bowlegged', and suggests that Sir Brian's disfigurement was an acquired rather than an inherited characteristic. As far as the contention, on the part of Dray and Scriven, that historians explain, and hence give explanations of events, Hempel might say only that he agrees, save that for the purposes of philosophical analysis he prefers to designate these as explanation sketches. Why quarrel over words?

So to some extent, the quarrel turns on verbal issues, but at the same time there is a different attitude towards history. Dray and Scriven appear to consider the practice of history perfectly satisfactory as it stands, but Hempel, if he does not openly advocate a reform of history in the direction of physics—the paradigm science—has certainly encouraged, among some historians at least, the view that history ought to be revised. Clearly he would not accept the thesis that such revision is impossible and that the distinction between history and any given natural science is ultimate, as the advocates of historicism have contended. Scriven's answer to this will be that the difference is indeed not ultimate, but that physics is *in fact* a good deal more like what history is *in fact*; that Hempel has misconstrued the logical structure of the paradigm science itself, and to this is due the mistaken notion that history requires revision—for there is no contrast of the presumed kind between history as such and physics as such, but between history and physics (or any science) on the one hand, and an idealized model, which corresponds to no actual science but only to a logician's fantasy, on the other.

But this in turn reflects a fundamentally different attitude towards the task of philosophy. Hempel might reply to Scriven's charge that it was not his purpose to *describe* science. That job is better left to the sociologist. Rather, he was seeking, as philosophers ought, rationally to reconstruct the concept of scientific explanation. His concern was to specify what conditions must be satisfied if we are to pretend to have adequate grounds for something of which an explanation is solicited. It seemed wholly plausible to him that we have adequate grounds for an explanandum when we are able to deduce it from premisses which satisfy a variety of conditions.[1] To the claim that general laws are part of the grounds of an *explanation*, as Scriven has insisted, and may be appealed to for purposes of *justifying* an explanation, he might ask how 'justification'[2] is to be understood? There can be no doubt that Scriven has widened the area of discussion by bringing in pragmatic factors, but is not the question of using laws to justify an explanation—the question of what he calls 'role justifying criteria'—the crucial issue? These 'role justifying criteria' are called into play when he who offers the explanation is required to answer the charge that perhaps 'no causal connection exists between the phenomenon as so far specified and its alleged effect'. But how is this charge to be met except by proving that the connection holds? What is this except to bring forth the appropriate general law? And once having done this, does the justification not simply take the form of a deduction? And in so far as it does not, to that degree the justification has not in fact been achieved. So if 'justification' does not entail deduction, the meaning of the former operation is far from clear.

So the tensions which existed among our original three propositions reappear as further tensions among the varying solutions to these tensions, these last growing out of competing attempts to fix the meaning of certain key terms in accordance with competing attitudes towards language, history, and philosophy itself. Yet, as with our original three propositions, there seems to be an element of truth in each of the positions offered in solution to the original difficulties. Once again, it is difficult to assent to any one of them if this requires wholesale rejection of those remaining. Now I think it is possible to cull this element of truth from all four conflicting positions, and to offer an analysis of the rôle of laws in historical explanation which might simultaneously satisfy all contenders; an analysis which will dissolve the difficulty by exhibiting all four as complementary rather than exclusive. Assuming this can be done, we can, I think, then turn to the problem of determining in what sense the form of historical explanations is that of narratives.

The four positions I have identified have mainly had to do with the structure of the *explanans*, and the moot question was whether general laws are to be included in *explanantia*. (*A*) and (*B*) said yes, but were divided about whether there are *historical* explanations. (*C*) and (*D*) sided with (*B*) on the existence of historical explanations, admitting there were such things, but said that no laws are included in explanantia. Whereas (*C*) said general laws are *in some sense* involved in explanations, (*D*) rejected this out of hand. Yet the issue which was not raised in any of the positions has to do, not with the anatomy of the *explanans*, but with that of the *explanandum*. I shall argue that there are *explananda* which logically presuppose general laws, and explananda which do not. Accordingly, whether or not there are to be general laws in the explanans depends upon our original description of the event for which explanation is sought. I shall argue, further, that if the original explanandum is not one which logically presupposes a general law, it can be replaced with one which does, and vice versa, so that the question of general laws is in some important sense connected with the question of how phenomena and events are to be *described*.

Here I should like to make an obvious and trivial point. Phenomena *as such* are not explained. It is only phenomena as *covered by a description* which are capable of explanation, and then, when we speak of explaining them, it must always be with reference to *that* description. So an explanation of a phenomenon must, in the nature of the case, be relativized to a description of that phenomenon. But then if we have explained a phenomenon *E*, as covered by a description *D*, it is always possible to find another description *D'* of *E*, under which *E* cannot be explained with the original explanation. If there are indefinitely many possible descriptions of a phenomenon, there may be indefinitely many possible different explanations of that phenomenon, and there may, indeed, be descriptions of that phenomenon under which it cannot be explained at all.

Unless we explicitly give the description, or unless an intended description of it is implicit from the context, there is no sense to be made of any request to explain the designated phenomenon. Here I shall be speaking of Hempel's analysis to begin with. Now my point is that, strictly speaking, I can no more explain or ask for an explanation of the Civil War than I can explain, or ask for an explanation of the piece of paper in the typewriter. Who would know what to make of the demand: explain that piece of paper!—unless a description were, so to speak, in the air, as for example, the piece of paper is white, or is here, or is blemished with jam spots? At best, the expressions 'the Civil War' or 'the piece of paper in the typewriter' are referring expressions. They can be the subjects of sentences, the objects of verbs, but by themselves they are not sentences, and hence, by themselves, are neither true nor false. Hence, for obvious reasons, they cannot serve as conclusions of deductive arguments. Briefly, they can only be covered by laws if first they are fitted into sentences, and the things they designate can then in principle only be covered by laws if first they are covered by descriptions. Linguistically naked, they are unintelligible. However, there are, in principle, descriptions which might cover them which logically prevent them from being covered by general laws.

Consider, for example, the admittedly questionable notion of a *complete* description of E. Let us imagine this to consist in taking all the logically discriminable true sentences about E, abstracting from each its predicative expression, and then conjunctively asserting all these predicative expressions of E, so that there will be as many predicative expressions in the conjunction as there are discriminable true sentences about E. It does not matter that we cannot furnish such a complete description. It is, however, by the principle of the identity of indiscernibles, logically impossible for two phenomena E and E', to have identical complete descriptions. But then under a complete description, E cannot be covered by a law, for such a law would be logically restricted to a single instance, and would accordingly be disqualified as a law.

But we need not restrict ourselves only to complete descriptions. By Hempel's own criteria for a general law, a proposed law L must contain 'no essential—i.e. uneliminable—occurrences of designations for particular objects'. Hence in so far as a description D of a phenomenon E contains such designations, it cannot, under D, be covered by a general law. It is a moot question whether all such designations are in principle eliminable. Supposing, however, a favourable answer were given to this question, so that in principle we might eliminate, from D, all such designations and produce another description D' which contains none; still, E cannot be covered by a general law under D—it being admitted that there are such descriptions—even though, under another description D', it *can* be so covered. But this is, as I shall show, a crucial consideration. It suggests that a phenomenon can be covered by a general law only insofar as we produce a description which contains no uneliminable designations of it. Or, briefly, we can cover an event with a general law only once we have covered it with a general *description*. But then it is easy to find descriptions of a phenomenon under which it cannot be so covered and cannot, accordingly, be explained along Hempelian lines. It immediately follows that any work of history will contain many descriptions of events under which the events cannot be explained if Hempel's model is correct. But it does *not* follow that the model is as such *incorrect* or that the events in question are unexplainable. Only unexplainable under the descriptions which have been given them. But then to explain these events requires a *redescription* of them. Indeed, to be able to redescribe the events is already, in a sense, to have explained them. For often we can only carry through the redescription when we know the explanation, and typically, again, the redescription entails a covering law.

But let us illustrate these claims by taking an example which is typical of a whole class of historical explanations, explanations which refer to the history of whatever it is that requires explanation. During the celebration of the last *fête nationale monagasque*, the streets were decorated, as one would expect, with the flag of Monaco. But side by side with these were to be found *American* flags. One might have wondered why, if there were American flags sharing pride of place with Monagasque flags, there were not flags of other nations, for instance English or French or German flags. Here is a context in which one feels the need of an explanation, an explanation indeed of two things: the presence of American and the absense of other flags alongside the national ones. And here is a wholly plausible explanation someone might give: he might tell us that the Prince married an American woman. At this point we might play Professor Dray's game; we might say that we know of no law which connects the event K (Prince Rainer III marrying the actress Grace Kelly), with the event E (Monagasques putting out American flags during their national holiday). Indeed, at this level of description, there is no law which connects these events, but with the appropriate *redescription* of each event, it is easy enough to furnish a law, and a law, in fact, which both licences and is licensed by the redescriptions. Furthermore, we can even, once having carried through the redescriptions, frame the explanation in deductive form.

Here is a triad of descriptions, at different levels, of the event E:

a. The Monagasques put out American flags side by side with Monagasque flags.
b. The Monagasques were honouring a sovereign of American birth.
c. The members of one nation were honouring a sovereign of a different national origin than they.

We may regard *a* as the description of the event before an explanation was available. We may continue to term it the *explanandum*. We may regard *b* as the description of the *same* event *after* the event has been explained. Had we known this description of the event to begin with, we would not have needed an explanation, either for the presence of American flags or for the absence of flags of other nations. We may regard *c* as a description of again the same event, though it could be a description of a good many different events of the same kind. We may regard *c*, indeed, as the result of eliminating terms designating particular objects in favour of general designatory terms which include the originally designated objects amongst their extensions. I shall term *c* the *explanatum*. Actually, *b* qualifies as an *explanatum* as well. The fact is that the move from *b* to *c* is relatively easy to make. The hard work is the move from *a* to *b*. It amounts almost to a transformation in perception, the objects, as it were, in the visual field remaining constant, but now seen in a whole new set of relationships. One has a genuine *sense* of having been illuminated. So one might yield to the philosophical penchant for making distinctions, and refer to *b* and *c* as, respectively, the *concrete* and the *abstract* explanatum. It is the latter which serves to put the event under a formal law.

But *what* law? I think it not difficult to state in at least a vague way what general law might be spontaneously advanced by anyone who felt illuminated by the redescriptive shift from *a* to *b*. It is something like

L. Whenever a nation has a sovereign of a different national origin than its own citizens, those citizens will, on the appropriate occasions, honour that sovereign in some acceptable fashion.

We assume, as independently known

K-1. The sovereign Princess of Monaco is of non-monagasque origin.
K-2. The *fête nationale monagasque* is an appropriate occasion for honouring sovereigns of Monaco.
K-3. Putting out the flags of a person's native country is an acceptable way of honouring that person as a native of that nation.

Having stated all these connections—and there might be persons for whom they should all have to be spelled out, as well as other connections, with special qualifications—there is no reason to doubt that we could, in the end, exhibit *c* as a deductive consequence of all of them together. Presumably the explanation here is the correct one, but it may take some doing to get it all stated in a correct formal way so that the deduction might go through. I have no objection, then, if someone were to say that I have provided only an explanation sketch. I have in a vague way indicated a law, even stated that law, and specified the relevant initial conditions. Nor have I any objection, if someone wants to point out that there is no clear *need* for working the explanation up, that all of what I have stated will perhaps be spontaneously understood by the person to whom the explanation is made. Pragmatically there is no point, but we are dealing with a philosophical question. Students of formal logic might find it equally tiresome to spell out all the steps required for a formal demonstration of an argument whose validity they intuitively recognize.

Now there are many questions left to discuss, but I should like to pause for a moment to indicate how the above analysis might be acceptable to the three last positions in the four specified earlier.

Surely, this analysis must satisfy Hempel. We have a law, which covers the event. We have given an explanation sketch which could, with appropriate care, be made into a fully-fledged explanation, and in general, apart from the remarks on explananda and explanata, the analysis is mainly his. On the other hand, it should really satisfy Dray as well. To begin with, the law *L* does *not* cover the event *E* as such, nor is *E* covered even under the description *a*—which is just the sort of description one would be most likely to find in history books. You cannot, moreover, deduce the explanandum (*a* again) from the explanans. It is what I have termed the *explanatum* which may be so deduced, and it may be some comfort to him to point out that we are able to provide the explanatum, and indeed the law, only after we have had the explanation given us (or found it out for ourselves). The replacement of the explanandum with the explanatum (a with b and thence c) ought not to strike him as illegitimate. It is, in fact, exactly an instance of what he has elsewhere termed 'explaining what' in history. In fairness to Dray, it might be said that the man who gave the explanation did not *use* the law L. But to someone unfamiliar with honouring customs, with national symbols, and with a whole related meshwork of general notions or concepts, the explanation would be utterly opaque. The explainer would be obliged explicitly to mention some such law as L in such a case to justify his explanation, and this brings us to Scriven. For L would not merely play that justificatory rôle he has so much emphasized, but it would qualify as a truism or, in his terms, a 'normic sentence'. Moreover it gives a sense to the notion of justification; the sense of permitting a deduction. This will be wholly compatible with the contextual and pragmatic considerations he has so carefully outlined.

So it seems to me that by paying some attention to the question of how phenomena are to be *described*, we can elaborate a theory of the rôle of general laws in history which will reconcile, details apart, the main arguments of three of the main positions. But we have yet to come to terms with A. To this I now turn.

Strictly speaking, the analysis just sketched must appear flatly incompatible with A, for it clearly supports Hempel's contention that the pattern of explanation is indifferent to the distinction between human and non-human phenomena, or any purported difference between the 'human' and the 'natural' sciences. There are, in fact, two considerations pertinent to my example which plainly show that we do *in practice* appreciate, and indeed understand human behaviour, in the light of very general concepts which, if stated explicitly, would take the form of general laws. The first is the plain fact that in having explained the putting out of American flags by referring to the national provenance of the female ruler of Monaco, most people, assuming them equipped with a modicum of relevant historical information, spontaneously 'see' the connection without having to fumble for the appropriate general law in virtue of which the connection holds. The very fact that they do not need to make explicit to themselves, or require others to make explicit for them, what this general law is, shows, or strongly suggests, that thinking in terms of general concepts is so natural as to be very nearly unconscious. It is accordingly very easy to see why, in view of the psychological fact that the laws in question are never consciously entertained (or seldom so), philosophers would be tempted to say there is no general law, or that no general law is required to understand the explanation. The second consideration is that the felt need for an explanation, the sense of puzzlement, frequently, even typically arises when we are *unable* to assimilate the phenomenon in question into a given general concept. We think: the flags are out to honour a friendly nation. But this does not fit. For why not the flags of other, indeed all, friendly nations? So it is with respect to some accepted general concept about flag-displaying behaviour that we are puzzled, and this again shows how prominent a rôle, even in the pragmatic aspects of the problem, general laws play. Someone, say a child, who was unconversant with the applicable general concepts, could only be equipped with a sense of puzzlement if he was first instructed in the appropriate generalities; lacking these, he feels no need for explanations. Meanwhile, it would be incorrect to say the events felt to be in need of explanation cannot be covered by a general law on the grounds that they are, to begin with, appreciated as *exceptions* to some general law. For it only follows that they are exceptions to a general law which it was a mistake to apply to them originally, and not that they are exceptions to *every* general law. There is, in explanation, an analogue to the phenomenon of illusion. Putting out flags *could* be an instance of honouring a friendly country. In this case it was not an instance of that, and the puzzle might have arisen from simply trying, and failing, to subsume it as an instance of that general concept. But the illusion is dispelled when we see the event, not as an exception to the general law originally proposed, but as a *proper* instance of a *different* general law altogether.

I shall return to these considerations in a moment. So far it seems that we have a strong counter-argument against one of the basic commitments of *A*. Yet it must be admitted that there remains something that can yet be done on behalf of that position. Let me begin by pointing out a further feature of Hempel's analysis. He claims, and has been severely criticized for doing so,[1] that explanation and prediction are of a piece, logically speaking; that we use exactly the same apparatus whether we explain an event which has happened or predict an event which will happen if our apparatus is satisfactory—the difference being only when, relative to the time of the event's occurrence, we put the apparatus into play. We can be said to have explained an event if, and only if, we could have successfully predicted the event before it happened, using exactly the same law. This claim would certainly be objected to by the adherants of A, one of whose main theses is that human beings are free,[1] for it might seem to them wholly incompatible with human freedom that human behaviour is predictable. But then, accepting Hempel's rule for converting explanations into predictions, and vice versa: if human behaviour is explainable, it is predictable. So, if they wish to say it is unpredictable, they will also then want to say it is unexplainable by means of the use of general laws. Whether or not they are right in connecting predictability with the negation of the thesis that humans are free is an issue I do not wish to probe at this point. None the less, I think it possible to give a finer analysis of the relationship between explanation and prediction which will render my analysis of general laws acceptable to them, and so salvage part, at least, of their position. Once again I shall pay some particular attention to the question of description.

Now it seems to me that there is no doubt but what, if we knew the law L, and knew, in addition, that the monagasque Princess was of non-monagasque origin, and that a *fête nationale* was an appropriate occasion for honouring sovereigns, we could, with some security, have predicted the event in question. Or rather, we could have predicted it under the description c or b—the description we earlier characterized as the explanatum. Yet it does not follow that we could, with the same security, and on the basis of the same information, have predicted the event under the description a. For though a, b, and c are all descriptions of the same event, and though c and b are deductive consequences of L, together with those initial conditions we have specified, a is *not*. I shall contend, now, that the law L—and a great many, if not all the laws which are elements of explanations in history—covers a class of instances which is both *open* and *non-homogeneous*. This is so because the descriptions which serve as *explanata* have open and non-homogeneous classes of events as their extensions. To refer to my illustration, there are many different events, many different things which the monagasques might have done, which would indifferently be covered by *c* (or *b*) and by *L*. Putting out flags is but one way of satisfying the general description. Given the appropriate conventions, the description *a* entails the description *c*, but *c* does *not* entail *a*. For, appealing to the same conventions again, *c* is compatible with not-*a*. One can honour a foreign-born sovereign without necessarily putting out the flags of that sovereign's native country. So one could have correctly predicted *c* without also predicting *a*; or one could have been correct in predicting *c* (or *b*) and have been *incorrect* in predicting *a*. For a different event, *d*, might have happened instead of *a*, and yet be covered by the same law *L* and the same general description *c*. It will not do, accordingly, to say, simply, that we can predict the *event*. The question is under what description the same event might have been predicted.

This point is connected with an earlier one, where I spoke of the construction of narratives on the basis of documentary and conceptual evidence. In seeking to fill a gap in historical knowledge, where documentary evidence ('history-as-record') is unavailable, we may tentatively, and on the basis of some general law of the sort I have been discussing, or on the basis of some type-concept, postulate the occurrence of an event, or set of events, to fill that gap. We might, by chance, have hit upon what in fact happened. But a type-concept is just what it is: it has to do with what typically can be the case, and this is compatible with a whole range of qualitatively different events, all of which satisfy the same general description, and any one of which *can* have happened. But to be in a position to know which event in the range *in fact* happened, we require documentary evidence. Conceptual evidence, accordingly, at best supports a *plausible* account.

The laws, then, which may be said to be implicit, according to Hempel's account, in typical historical explanations, are peculiarly loose, in the sense that they can accommodate any number of qualitatively different instances. They indeed permit *creative opportunities*, for the class of events they cover is open, in the sense that we can in principle always imagine an instance, covered by them, which need not in any obvious way resemble past instances. It is this sort of situation, for example, which allows us to class, as works of art, things which do not necessarily resemble objects already classed as such, and which permits artists to pursue novelty which, should they succeed in finding it, does not automatically disqualify them from having produced a work of art. What we have to do with here, in fact, is what Wittgenstein has notoriously discussed under the name 'family-resemblances'. It is in the nature of families to produce members of themselves whose exact resemblance to any existing member is more frequently the exception than the rule. The laws we appeal to, then, in historical explanations, could be inserted in, or made to justify, a practically inexhaustible class of explanations, none of which, as Dray has properly contended, actually *entails* the law. On the other hand, they may be said to entail the law when taken in conjunction with what I might term *rules of re-description*, in accordance with which we may replace a given description of an event with one of greater generality. This set of rules is perhaps difficult to specify; what, for example, are the criteria by which we class something as a work of art? It is, moreover, not always a simple matter to effect this re-description, for the same event may *appear* to sustain one general description when in fact it requires a different one. It is this which makes logical room for what I have called *illusions of explanation*.

Now there is one obvious objection which might be raised against this characterization of general laws. Suppose, in accordance with my analysis, one has explained an event, that is, one has covered it with a general law after first covering it with a more general description; one has made the move from explanandum to explanatum. The fact remains that exactly the same explanatum and general law could have held though the event explained was quite another one from what in fact occurred; was described with a different explanandum than the one we gave, so that the identical explanatory apparatus could hold whether the explained event occurred or not. Thus let E be the event in question, and let D be the explanatum of E. But suppose that E' is an event qualitatively distinct from E but nonetheless also covered by the general description D. The question now is why, since our explanatory apparatus cannot discriminate amongst the instances which it covers, we want to say that we have explained E, for we might, with the same apparatus, have explained E'. Suppose someone claims that we really have not then explained why E happened instead of E'. For the description D does not entail E any more than it entails E'—though either of these, with the appropriate rules, will entail D. Since the explanation is compatible with other possible happenings, are we entitled to regard it as an explanation at all? As Alan Donagan writes:

If it is supposed that an explanation need not logically entail what it explains but may be consistent with several other possibilities, then it will fail to explain why one or another of these possibilities was not realized, will fail to explain why what it purports to explain should have happened rather than something else.[1]

It is interesting that Donagan's statement here was offered by him as one of the arguments *in favour* of the Hempelian model, but that we are actually able to raise it as an objection *against* what is essentially a variant of the Hempelian model. On the other hand, it is not quite true to say that the explanation does not logically entail what it purports to explain. It *does* entail the explanatum. The problem is that it does not entail one rather than another of the various instances indifferently covered by the same explanatum. Thus we have to make a distinction between two senses of 'other possibilities'.

To begin with, there is one sense in which our explanation does rule out other possibilities. Consider only the action of putting out American flags with Monagasque flags during the *fête nationale*. This *could* be covered by any number of different general descriptions or explanata. It could, for instance, be an instance of the description 'honouring a foreign power'. It could, indeed, be an instance of the general description 'insulting a foreign nation': putting out American flags might constitute a studied insult against France. It could be both together. But these general descriptions are ruled out as 'other possibilities' by our explanation, assumed to be correct. It is perhaps a little strong to say they are ruled out, for there is always the possibility of overdetermination. Even if we accept the Principle of Sufficient Reason, and say that for each thing that happens, a sufficient condition for its occurrence must have obtained, this does not preclude the possibility of more than one sufficient condition having obtained. Nor is it an argument against something being an explanation that there should be another explanation of the same thing. People do kill two birds with one stone, go walking to get exercise and lambchops, and the Monagasques could at once have been honouring their sovereign, honouring one foreign country and insulting another, and do all this by performing just the one action. But these are complications we may momentarily overlook, and simply say that, except for overdetermination, other possible general descriptions are ruled out when we have succeeded in covering an event with the correct explanatum. Having done this, however, there remains the second sense of 'other possibilities', namely, those other co-possible events, covered by particular descriptions each of which satisfies the same general description. And since the latter fails to discriminate amongst these, it does not rule out any of them. We could not deduce which of them held if we only knew that the general description were true. It is the sort of thing we can know only by independent historical investigation, and inasmuch, I add, as there is no comparable historical investigation for the *future*, our knowledge of the future is general and abstract by comparison with our knowledge of the past. Knowing even all the general laws there might be, we could not, if the laws were the sort I have been speaking of, predict the events of the future under particular descriptions. The present objection, however, is that we cannot *explain* events under these descriptions either.

We have, then, these two senses of 'other possibilities'. The first is that the same event may be covered by different general descriptions. It is this fact which makes overdetermination possible as well as illusions of explanation. The second is that the same general description can cover qualitatively distinct events. This fact serves to account for the general character of our knowledge of the future, and, symmetrically, for the further fact that without documentary evidence (history-as-record) our knowledge of the past would be as general and abstract as our knowledge of the future. Our explanation serves (discounting overdetermination) to indicate which, of the classes our event *might* be an instance of, it is *in fact* an instance of. I shall say that it rules out other possibilities in the *class-sense*. But having isolated the class the event belongs to, it does not allow us to say which of the many possible instances which could have held, did *in fact* hold. I shall say it fails to rule out other possibilities in the *membership-sense*. This would perhaps be unimportant if the members of the class in question were homogeneous, but since they are not, this failure *may* be regarded as a defect in this theory of explanation.

Now I am not certain that I can satisfactorily meet this objection.[1]

Nonetheless, a few things can be said. First, the objection presupposes the existence of other possibilities in the membership-sense. If there were no such, we could hardly be held responsible for having failed to elaborate a theory of explanation which accounts for them. But since other possibilities, in the membership-sense, are to be specified relative to a general description and a general law, of which they are instances, the objection again presupposes just the characterization of general descriptions and general laws which I have offered. So in effect the apparatus I have elaborated is tacitly accepted by the objection, and is, accordingly, assigned at least a partial validity. Secondly, let us suppose that we could find a description of some intermediate degree of generality between a and c (or b), relative to which some of what heretofore have been other possibilities in the membership-sense are converted into other possibilities in the class-sense. Even so, the fresh explanatum, providing it remained of the kind I have been discussing, would leave us with the same general difficulties, if difficulties they are, for there would still be *some* other possibilities in the membership-sense. Indeed these difficulties would dog us all down the line, until we reached some kind of description of the event relative to which there were *no* other possibilities in the membership-sense. These descriptions might be of two sorts. Either they would be descriptions like a, or they would be general descriptions with homogeneous classes of instances. If the first, we are left with the question of a *general law* which will have a as a deductive consequence, and, by Hempel's criterion, since the law would contain designations of specific objects, it would disqualify as a general law automatically. If the second, it would be hard indeed to state what could be the law, and almost certainly false to say that such a law is presupposed in historical explanations, because any such law is almost certain to have non-homogeneous and open classes of instances. It is debatable whether we ever shall have such laws, and I can see no way of deciding, *a priori*, how such a debate could be resolved.

Meanwhile, in the nature of the case, it is not a simple matter always to say in advance what the other possibilities in the membership sense are; it is particularly difficult to specify the entire membership of the class. Perhaps it is impossible to do so, for there is always the possibility that human inventiveness will contrive a novel instance which we can recognize afterwards as belonging to the class but which we could not have anticipated even though, in a general way, we might have predicted the general description this instance falls under. In a comparable way, even knowing that a man has a disposition to do kind things, and knowing that a given occasion is one on which he can be expected to do something kind, it is not always a simple matter to say what precise kind thing he will do. To be kind is to be creative in benignity, to be considerate, to surprise people by the singular appropriateness of one's gestures. To ascribe such a disposition to a person is then to allow room for creativity, kindness not being a ritual affair, and there being no precisely enumerable set of things which exhausts the manner in which the disposition functions. Many character traits are of this sort, so that one frequent claim of determinists, that if we but knew precise information about a person's character and about the conditions he was operating under, we could infallibly predict his behaviour, is only half right. We could do so under a general description, to be sure: we can safely say that the kind man will do the kind thing, the witty man say the witty thing, but this does not mean that in a more specific way we can predict the former's displays of good-heartedness of the latter's quips and *bons mots*. We can recognize them afterwards as proper instances without being able to predict them.

But in view of these considerations, it is somewhat empty to demand an explanation of why E happened rather than something else co-possible with E in the membership sense. One can, to be sure, perhaps specify some possible event E', and ask why E' did not take place while E did, and point out that E' would have served the purpose as well as E. But the explanation for the non-occurrence of E', while it may take various forms—to do E' never occurred to the individuals, or occurred to them and was rejected for this reason or that—it need not in principle require laws different from the sort we have considered, would have to be made case by case, but would, in the end, be liable to exactly the same objections. So pending the provision of another sort of law altogether, the objection reveals a feature inherent in the structure of explanation and basically favourable to position A.

Perhaps too favourable, it might be argued. For the example I have chosen might seem specially selected to put A in the best light. This is not quite true. It was selected as a typical instance of a kind of historical explanation, and in fact was found to satisfy the other positions first, one of which was diametrically opposed to *A*. There can, on the other hand, be no question but what, in some sense of the term, we should only be able to decide which general description to give of the action—putting out flags—by verifying some fact about the Monagasque mind, and what their precise intentions were. Yet it does not follow that we need perform some operation of empathic projection to verify any such fact. To begin with, the emphasis I have placed upon creativity is already a limitation on this notion. For we ourselves, when we behave creatively, often find that we have hit upon a certain thing without being clear how we did it, or what, if anything, went on in our minds at the creative instant. To identify empathically with our minds at that moment would leave the empathizer no clearer than we, assuming he was successful in simulating our own mental state. Secondly, there are clearly general descriptions of actions under which the action was not intentional, in which case, of course, empathic projection would naturally be inappropriate. Finally, and as a special case, the general description—the explanatum—which the agents themselves would place upon the event might, on the basis of historical knowledge, turn out to have been the wrong one, an illusion of explanation which only later historical research is able to rectify. But in that case the explanation of the event under this new, and presumably correct description, would be far different from the one which those involved in the event would have given. So the participants in the events, as my chapter on narrative sentence has, I hope, demonstrated, have no privileged status when it comes to historical explanations.

I have tried, thus, to salvage that part of the truth which each of the positions arrayed can be said to have, and to synthesize a theory of the rôle, in historical explanation, of general laws; a synthesis which would satisfy all four positions as well as accord with actual practice. But where is there room in all of this for the position I committed myself to defend? For the theory I have synthesized seems, on the face of it, complete. What need we now do, in explaining events, save cover them with general descriptions and thence general laws? What further point need be made? What need have we for narratives? To these questions I shall devote the next chapter.

XI

HISTORICAL EXPLANATION:
THE RÔLE OF NARRATIVES

The only support so far given to the claim that stories, or narratives, are forms of explanation, is the pragmatic consideration that in certain contexts what people typically want and expect, when the need for explanation is felt, *is* simply a true story. So much is a factual matter, and presumably beyond controversy. Beyond controversy, moreover, is the fact that the need is typically eased only when a story of the required kind is provided, and the further fact that those who are called upon to explain something will naturally tell a story. But what conditions must be satisfied by such a story remains to be determined.

I shall begin by once more looking to the description of that for which explanation is sought and given. One thing which, while it has been remarked on in discussions of explanations in history and elsewhere, has not, I think, been sufficiently appreciated, is that the explanandum describes not simply an event—something that happens—but a *change*. Indeed the existence of a change is often built into the language we employ to describe things: the description makes an implicit reference to a *past* state of the subject of change. I have already referred to the use of temporal language in my discussion in chapter V. Here, once again, we find that implicit reference to a lapse of time is already incorporated in explananda. Simply to describe an automobile as *dented*, for example, is implicitly to refer to an earlier state of this same automobile in which it was *not* dented; and to demand an explanation of the dent is accordingly to demand an explanation of the change. We require, of stories, that they have a beginning, a middle, and an end. An explanation then consists in filling in the middle between the temporal end-points of a change. The chief difficulty in regarding S, in chapter VII, as a story, is that there seemed, on the face of it, no connection between the temporally ordered events mentioned in it. No later event mentioned seemed, in any

233

obvious way, to refer to any earlier event also mentioned in *S*, and hence no intermediate event mentioned in *S* stands as a middle between the events which flank it temporally. *S* then consists in a sequence of beginnings or endings, but not beginnings or endings of the same stories. Or perhaps the events it mentions are middles in stories the beginnings and endings of which failed to get included in *S*. A story is an account, I shall say an explanation, of how the change from beginning to end took place, and both the beginning *and* the end are part of the explanandum.

Consider now two examples studied by recent philosophers of history, Professor Gardiner's[1] (and Professor Dray's) example of Louis XVI dying unpopular, and Professor Nagel's example of a change of attitude on the part of the Duke of Buckingham.[2] To say that Louis XVI died unpopular is presumably to presuppose that Louis was not always unpopular, for then his unpopularity could hardly be explained with reference to policies pursued by him which were felt to be detrimental to the French national interests. Reference to these then serves to explain the *change* in attitude towards that king. It roughly constitutes the middle in the story of how people's attitudes towards Louis changed. The beginning and the end of the story are the end-points of the change, and belong equally to the explanandum.

Again, and in an obvious way, when Nagel speaks of explaining the opposition, on the Duke of Buckingham's part, to the proposed marriage between Prince Charles and the Infanta, the presumption is that the Duke was not *always* opposed (for there would then be no tale to tell), but that there was a definite *change* in the Duke's feelings towards the marriage. But it is a mistake to say *simply* that what we want an explanation of is the Duke's opposition to the marriage, and to provide 'The Duke was opposed to the marriage at *t-1*' as the explanandum. What we want to have explained is the shift, and a more appropriate explanandum would be a *narrative sentence*, one which refers to two distinct events, for example (to use Nagel's own formulation) 'Buckingham changed his mind about the desirability of the marriage, and became an opponent of the plan'. It is important to note the temporal vocabulary in this candidate explanandum. The Duke *changed his mind*, the Duke *became* an opponent —implying that he was either neutral or a proponent before. From this it follows that it is a mistake to regard the earlier event referred to as part of the explanans. For this is to mislocate it on a logical map of the structure of historical explanation. We could, indeed, describe the earlier event with a narrative sentence which referred to the later event, that is, not simply with 'The Duke favoured the marriage at t-0' but with something like 'The Duke, who was later to oppose the marriage, was until early in 1623 a supporter of the alliance'. It is a matter of indifference whether we say that we want to explain the later event, or the earlier event under the narrative description, for it is the *connection between* the events which has to be explained.

This connection is *not* a causal connection: rather, the events in question are connected as end-points of a temporally extended change—as the beginning and end of a temporal whole—and it is the change thus indicated for which a cause is sought. It seems to me then that Nagel misreads the connection, for he remarks that it is 'difficult to imagine a reasonable generalization which would permit us, given c_0 [Buckingham desires the marriage between Charles and the Infanta] to conclude that c_{12} [Buckingham changes his mind] would probably occur'.[1] And he says that 'there appears to be no connection between c_0 and c_{12} (the action for which an explanation is being proposed) other than that the latter is the "opposite" of the former'. But there is a connection, and Nagel has in fact already stated what it is. It is only that he was seeking a different sort of connection. The connection is that of part to whole. The earlier event is part of what has to be accounted for, and reference to it is already contained in the description 'The Duke changed his mind'. If this is so, then it would be a clear case of begging the question to suppose that the earlier event belongs to the explaining apparatus, used to account for the change. It is part simply of the change, and accordingly part of what has to be explained.

Now to speak of a change is implicitly to suppose some continuous identity in the subject of change. Traditionally, indeed, it was felt to be a metaphysical necessity that some unchanging substance must endure through a change, it being otherwise a misnomer to speak of change at all. Without pausing here to worry about substances, we must still speak of the subject of change, whatever metaphysical status the subject is to enjoy. Hence, to keep to our examples, and pending any further analysis of what is involved, the Gardiner–Dray example has to do with a change in attitude, on the part of the French people, towards their king: '*they*' changed their attitude. The Nagel example has to do with a change of mind on the part of the Duke of Buckingham: *he* changed his mind. It is this implicit reference to a continuous subject which gives a measure of unity to an historical narrative. This gives us a further reason for dismissing S as a narrative: it was never *about* the same thing. Therefore since there was no subject, there was, strictly speaking, no change.

The *form* of an explanandum in history may be represented as follows:

$E:$ x is F at t-1 and x is G at t-3.

F and G are predicate variables to be replaced, respectively, with contrary predicates; and x is an individual variable to be replaced with a singular referring expression which designates the subject of change. Thus we get

$E:$ The Duke of Buckingham favours the marriage at t-1 and the Duke of Buckingham opposes the marriage at t-3.

The shift F–G is the change in x which requires explanation. But to explain the shift requires reference to *something happening to x at t-2*, an event (of whatever degree of complexity) which *caused* the change in x. I therefore offer the following model as representing the structure of a narrative explanation:

(1) x is F at t-1.
(2) H happens to x at t-2.
(3) x is G at t-3.

(1) and (3) together constitute the explanandum, (2) is the explanans. To furnish (2) is to explain (1)–(3). Without worrying for the moment about the question of general laws, I wish to point out that it ought now to be perfectly clear in what sense an historical explanation takes the form of a narrative. It is so in the sense that (1), (2), and (3) simply has already the structure of a story. It has a beginning (1), a middle (2), and an end (3).

It might be objected at this point that my model, if it may be so designated, is in fact satisfied by *any* causal account whatsoever. We can easily fit it to Hume's paradigm, for example: Billiard ball A is stationary at t-1, is struck by billiard ball B at t-2, and moves at t-3. But if this is so, the objection continues, my analysis fails satisfactorily to differentiate between historical explanations and causal explanations in general. I do not regard this as a very damaging criticism, however. For I should be perfectly content if I had shown both that there is no intrinsic difference between historical and causal explanations, and that causal explanations do in fact all have the form of stories. It may, of course, be true that there are explanations in science which do not have the narrative form. For instance, if we think of a physical system all of whose states are, in the appropriate sense, determined by an arbitrarily chosen initial state of the system, where the explanation of the system being in a given state consists in deducing the values of this state, in accordance with certain algorithms, from the values of the variables of the system at the initial state. But it should be noticed, as Russell pointed out, that the notion of *cause* has no place in such a representation.[1] I am interested only in causal explanations.

Secondly, it may be objected that any such explanation can always, in principle, be reconstructed in such a manner as to yield a deductive argument. This may be correct, but it would at best constitute a *formal* difference; a different way of expressing an explanation. This would leave intact my own claim that a narrative is a *form* of explanation. Meanwhile, it is worth noting that Hempel, who has been identified as the most uncompromising champion of the deductive argument model, presents his own celebrated example of the burst radiator in what is unmistakably a narrative form.[2] So it seems to me that there is as much justification for the claim that we can reconstruct a 'scientific explanation' as a narrative as there is for the reverse claim, and that an account in narrative form will not lose any of the explanatory force of the original, assuming it had any explanatory force to begin with.

Parenthetically, there is a certain resemblance between the narrative model and the alleged *dialectical* pattern which Hegel famously contended is exhibited everywhere throughout history. To some extent this is merely a numerical coincidence: thesis, antithesis, and synthesis can be mapped on to the structure of beginning, middle and end. Indeed, by applying narrative descriptions, we could almost give a sense to the Hegelian claim that the thesis 'contains' the antithesis and the synthesis. It is difficult, however, to know how far this analogy could be pressed, and I shall not attempt to elaborate upon it here. I turn, therefore, to the question of the place of general laws in narrative explanations.

It seems to me beyond argument that any decision one makes about what is to constitute the crucial middle in a narrative, the event *H* (that which happens to *x* and which causes *x* to change) must be selected in the light of some general concept, expressible, perhaps, as a general law. *H* must be the kind of event which can produce a change of the kind *F–G* in the subject *x*. One might at this point remark that narratives, rather than being simply explanation sketches which mark the place where laws are to be inserted, might instead be regarded as the *result* of taking an explanation sketch which makes use of general laws already, these marking the place where the description of an *event* is to be inserted. That is, where we are certain of the law but uncertain as to what precisely happened, the narrative then consisting in an account in which the *general* knowledge of what *kind* of thing must have happened is replaced by the specific knowledge of what specific thing, of the required kind, *in fact* occurred. This is far closer to the idea of a sketch than any actual piece of first-rate historical narration is, which seems, on the face of it, not to be like a sketch at all, but instead to be complete and finished.

Suppose, to illustrate with a simple case, we know that a change has taken place in a certain *x*—say an automobile—between *t*-1 and *t*-3. The change consists in the shape of a bumper: it is dented where it was not dented before. Now we wish to account for this change. *A priori*, we normally assume some such deterministic maxim as this: automobiles do not just spontaneously change in the fashion remarked upon: they only do so when something happens to them. This is hardly a law. It is, better, a methodological directive which assures us there is a story to be told, sending us to look for a causal episode. But we have more than this directive to go on: we will normally be able to specify in a general way what sort of causal episode must have occurred. And this is to appeal to a general concept which allows us to *posit* an explanatory middle to the story the beginning and ending of which we know. We can do no more than posit a causal episode under some such general description as 'something *y* struck *x* with a certain force at *t*-2'. Just on general principles, then, we can be said to know the truth of the following narrative:

I.   The car is undented at *t*-1.
II. The car is struck by *y* at *t*-2.
III. The car is dented at *t*-3.

This, however, is really just an explanation *sketch* for a narrative, a narrative only being available when we know what hit the car when. So *II* marks the place where, in the light of a known general law, the description of a particular event is to be inserted to convert the explanation sketch into a fully-fledged (narrative) explanation. This required description can be found only by making an historical investigation,[1] *guided*, no doubt, by the narrative sketch and the general law in accordance with which it was constructed. For there is no way, apart from such an investigation, of determining what was the specific instance covered by the known general law and the known general description. From these alone, for instance, we could not deduce

> *II'*.   A truck struck the car at 3.30

—for the narrative sketch could be true while the narrative consisting in replacing *II* with *II'* be false. Yet *II'* is the *kind* of event which could complete the story and explain the change (tell what happened and explain what happened at once). Yet, on the basis of the identical general principles which allow us to know that the narrative sketch is true and that the narrative *could* be true were *II'* to replace *II*, we can also know that the following does *not* complete the story nor furnish an explanation of the change:

> *II"*.   The driver of the car coughed at 3.20

—even though the event thus designated occurred and did so within the interval defined by t-1 and t-3. I do not say that *II"* might not ultimately need to be mentioned in the story: the driver's cough may have been violent, may have distracted his attention from the conditions of the road and exposed him to collisions, but these are complexities I need not grapple with here. For the only point I am seeking to make is that the construction of a narrative requires, as does the acceptance of a narrative as *explanatory*, the use of general laws. But these must, as we have seen, be supplemented by rules which allow us to identify the things that happen as instances of the general description which is as much as the general law allows us to give. Thus not merely does *II'* satisfy the general description in my narrative sketch, but so does

> *II'''*.   The owner of the car struck it with a sledge-hammer at 3.30.

The general law, no more than the general description *II*, cannot tell us which of these is the case. For this we require what I have termed 'documentary evidence'. And this is a job for historical inquiry. On the other hand, *II''*, while compatible with either account, could more readily be seen as belonging to a narrative completed by *II'* than one completed with *II'''*.

The problem is somewhat more aggravated when it is a question of explaining the shift in attitude towards the marriage of Charles on the part of the Duke of Buckingham. We can say, indeed, that something must have caused him to change his mind, but this is hardly more than to insist upon the methodological directive to *look* for a cause. Moreover, our general knowledge of automobiles permits us to say, without further concern, that if there is a dent, it must have been struck by something with a certain force. It is not so simple to say what sort of thing could cause men in general to change their minds about marriages. Here we would have to know what sort of man the Duke of Buckingham was. Suppose even that we know that he was a proud man, however, we would be able, at best, to offer, as a general description of the 'middle of the story', that something caused this disposition to actualize in a change of attitude. Yet this would still leave open a variety of *other possibilities* of the class-kind. The Duke, for instance, may have been an astute politician, and have seen an alliance with the Hapsburgs of Spain as detrimental to the national interests of England. He may have had personal ambitions, wanting to ally the Prince with someone from whose union he stood to gain. So we have a double problem on our hands. First to rule out other possibilities of the class-kind—something which is done almost as a matter of course in the case of the dented car—and then to rule out other possibilities of the membership kind. Not knowing what law is involved, historical inquiry is to that degree unguided. Once, however, we have the explanation, it is not difficult to find the required general description and the law. The explanation is given in the following brief account:

[King James's] son and his favourite Buckingham, indignant at their reception in Spain whither they had gone to hasten the negotiations, returned to England and declared themselves unwilling to participate further in the unholy alliance.[1]

This is a less detailed account, to be sure, than the one in Trevelyan's *England under the Stuarts*.[1] But C. V. Wedgewood is working on a wider canvas. The change in the Duke's attitude—a story in itself—is part of the middle in a larger story: the change in marriage plans of Charles. This, in turn, is part of the middle of a yet larger story, the change in policy on the part of James towards the Palatine Elector. This, in turn, is part of the middle in a further story, the change in England's position towards the war between Protestants and Catholics in Germany, and this in another story, the change in status of the Hapsburgs, the change in status of the Catholic Church . . . and so on.

Each of these changes is contained as part of the story of the next change mentioned, the final story containing them all. But then to explain the final story we must work our way back, step by step, to the Duke of Buckingham's shift in attitude. Changes are nested within changes, and stories require increasingly complex middles to explain the outermost change. We might represent it graphically as follows:

$$(\,)$$
$$((\,))$$
$$(((\,)))$$
$$((((\,))))$$
$$\ldots (((((\,))))) \ldots$$

obviously, this is too neat. For we have cases like the following:

$$((\,) \, (\,) \, (\,))$$

which raise special questions of multiple causation and overdetermination, as well as cases like this

$$((\,)(\, (\,))0 \,)$$

where we have overlaps. But barring these complications, which are in the end problems in the concept of causality, I think no special further problem is raised by them concerning the relationship between narratives and general laws. Philosophically, this is as far as we need go in the matter. Yet I should like to add to this account a few words on the concept of causality.

First of all, it does not seem to me that we require, for the analysis of causality in history, any different analysis from the classical one given by Hume. What is involved in any of the cases I have considered is never more, I believe, than a constant conjunction of like events with like events. It is of course true that we ourselves need not have personally built up these associations by performing, individually, the inductive generalizations which permit us to make the causal explanations. There is a social inheritance here, and the bulk of the generalizations we employ have been built up over the generations and have been built into the concepts we most of us employ most of the time in organizing experience and explaining how things happen. So immediate, for the most part, is our descriptive response when confronted with an instance which falls under a concept, that it is easy to see how some philosophers should have been persuaded that a special sort of capacity, an intuitive understanding or *Verstehen*, is exercised, as though we knew immediately and directly, or could know immediately and directly, what caused what happening in this change or that in this subject or that. And of course we do, in a way. Especially when human behaviour is involved, for we are far more familiar, most of us, with how humans behave than we are with how any other group of things behave. But if our familiarity were to be brought to this level with *any* kind of behaviour on the part of any kind of thing—animals, machines, or electrons—we should respond, here too, with the same immediacy and certitude.[1] Philosophers might argue that no law is needed nor used in the explanation of such behaviour—the explanation is just 'seen'. But psychologically true as this may be, it leaves unaffected the logical features of explanation, and should something happen of a wholly unprecedented sort, totally unlike any other change which ever took place so far as we knew, even in the case of human behaviour, we should, I submit, be totally unable to even begin to provide an explanation, to 'see' what were the causes of this change, until we had managed to cover the event with a general description and bring it together with like instances. If this should be doubted I can only, in the spirit of Hume, request him whose doubt it is to produce an instance, even an imaginary one, so remotely unlike anything ever seen before as to utterly defy this natural operation of the human understanding. In all the annals of history, surely, no such wildly deviant occurrence is to be found.

Yet, in acknowledging the basic cogency of Hume's analysis, I must enter some qualifications which my discussion of historical explanation seems to me to require. These, while going beyond Hume's account, are not in any sense to be regarded as incompatible with it. If anything, they extend and amplify his account and help to make clear why so frequently people have felt that in history Hume's analysis meets its nemesis, made up, as it is, of unique events and unprecedented causal sequences.

First, to accept the idea that like events have like causes, and that to speak of causes is to speak of constant conjunctions, requires nonetheless a stipulation that the likeness holds only at a certain level of generality. Our task, in cases where explanation is sought, is to find the correct general description of the event in question, to see it in the proper causal perspective. When we have achieved this, to quote the apposite law is easy, and nearly automatic. For we know, albeit still in a general way, what must have been the *kind* of cause which was responsible for the change. But there is a considerable distance between the establishment of this connection, and the identification of the specific event which falls under the general description. In history this is particularly so, for there is an endless variety of instances which fall under roughly the same general description. Indeed, part of the fascination of history lies in this spectacle of an innumerable variety of qualitatively different actions and passions, exhibited by human beings down the ages, which are for all that still instances of the same general description, and are covered by the same general principles we employ in everyday life, principles which, if enunciated, come in the end to be little more than truisms. It is for this reason, again, that we learn very little from history in the way of fresh general principles which we have not already acquired as part of our cultural inheritance. It is this, in turn, which supports the frequent claim that history is not a science. And if one of the things we expect from a science is the discovery of new general principles, then this charge is almost certainly true. This does not, of course, prevent us, if we wish, from collecting all these truisms—which have, perhaps, more verifying instances than any scientific law—into a body and labelling this a science. But this would require us to speak of common sense as a science, and would only cloud some relevant distinctions. On the other hand, these magnificently supported general principles, abetted by even the most extravagant imagination, would never have enabled us to predict the immense variety with which these principles have been illustrated and exemplified in the past.

Secondly, it is easy to see why there should be felt a certain looseness in causal explanations in history; why we fail to feel, in history, the sort of inevitability between cause and effect that we believe we ought. Hume brilliantly analysed the psychological origins of the concept of causal necessity, arguing that it is not found in events, is not objective, but is read into the conjunctions of events said, respectively, to be causes and effects, and is a habit of mind. But an exactly comparable psychological explanation can take care of the feeling of *non*-necessity between events said to be cause and effect in history. It consists in the fact that the necessity only holds at a level of generality in terms of which we do not usually think when we appreciate historical occurrences. Hume's billiard balls were notably and obviously like any billiard balls, it requiring a special effort to see differences between any chosen pair of billiard balls. Were a man able to pick, from a pile of billiard balls, just the one billiard ball he had played with a year ago in Pawtucket, he would exhibit a power of discrimination unshared by the majority of his fellows, who persist in treating such entities with a certain anonymity. For practical purposes they are right to so treat them, though *a priori* we know the differences are there. What is true for billiard balls is *a fortiori* true for collisions between them. But the instances covered by general descriptions—for example 'a revolution'—are obviously and often immensely different. We do not automatically think of just any revolution when we think of the French Revolution. Dents in automobiles are again causally homogeneous. We have a fairly circumscribed idea of what must have been the kind of thing which happened to a car in order to put a dent in it. But the sort of thing which might make a man like the Duke of Buckingham change his mind about the marriage of a Prince are not so easy to enumerate in advance. Once we know what turned the trick, we can bring it under a general principle readily enough. But at the same time, that very general principle admits of so many, and so various a set of instances, that we see no reason why this rather than that should have caused the Duke to change his mind, and so we feel less of the necessity, the certainty, than we do in the case of billiard balls or dented vehicles. Yet, at a certain level of generality, there is no difference between these cases. We may say that had the Duke been less arrogant, or the Spanish more ready to tolerate his behaviour, he would not have changed his mind. This may be so. But if billiard ball B had had a lesser mass, or if billiard ball A had been, by reason of being glued down, more resistant to impact, A would not have moved. The Duke's disenchantment with Spain, together with popular feeling in England against the Spanish Hapsburgs, together caused James to change his policy. But we could at the moment of impact tilt the billiard table, and say both caused A to move. The counterfactual problems are invariant.

That the explanandum, in a typical historical explanation, should be a description of a change, or imply a description of change, follows, I think, from the fact that what is being sought for and ideally provided by such explanations is a *cause* (or set of causes). For suppose we cite something y as the cause of x being G at t-n. It is clearly not enough simply to be able to demonstrate that y happened, that x is G afterwards (at t-n), and that y-like things can or regularly *do* cause x-like things to be G, so that there is a law-like connection between y-like things happening and x-like things being G. For all of this may in fact be true and still y may not have caused x to be G. Suppose, for example, that x is a female mammal and that x is pregnant at t-n, and that y refers to an episode of intercourse between x and some specific male mammal m at some appropriate time before t-n. There can be no question that such an event is just exactly the sort of event which causes x's to be G. Yet it by no means follows that x having had intercourse with m caused x to be pregnant, for it might have been done on a different occasion with a different male mammal. So we could, in principle, specify a set of laws and conditions which wholly satisfy Hempel's model, and which yet fail to explain x's G-ness. There would be one piece of information which Hempel has failed to make room for, namely information having to do with x's condition *before y*. Thus, suppose y takes place at t-2 and x is G at t-3, and y *can* cause x to be G. Still, if x was G at t-1, i.e. before as well as after y, then clearly y is not the cause of x being G. That which leaves something in the same condition it was before cannot be the cause of that condition in x. A cause must make a difference. Hence, from the fact that we seek for a cause when we try to explain something being in a certain condition, it follows that the explanandum, if only implicitly, is a description of a *change*. If the radiator was burst *before* the cold spell, the cold spell cannot explain the radiator's being burst.

I labour this wholly obvious point because it is not always obvious that we are in fact referring to a change when we demand an explanation of some event, and because it may seem unduly artificial to render explananda in terms of beginnings and endings. Consider, for example, my own illustration of the putting out of American flags during their *fête nationale* by the celebrating monagasques. It may plausibly be argued that all we are looking for in such a case is just an explanation of this event, and that we do not, so to speak, perceive the event as the terminus of a change. To be sure, we might be aware of some change or other. Thus: on Friday Monaco is flagless, a fact which we would hardly remark upon that day any more than we would remark that there were no eagles in Monte Carlo on Friday. But on Saturday the town is beflagged. The explanation of *this* change is that this is the *fête nationale*, but this is not the change that interests us and indeed it is being contended that we are not in any sense at all interested in a change, but only in there being American flags on the streets. Is it at all necessary to see this as part of a change and what, really, could we offer as the beginning of it? I answer that the explanation of the event *entails* a specification of what the beginning was, for it tells us what sort of change took place. For we have cited the marriage of the Prince to an American-born woman as the explanation of this event, and have now perceived the event under the general description 'honouring a foreign-born sovereign'. Now surely, if this *is* the explanation of the event, then putting out American flags was not something done on *fêtes nationales* by monagasques before the marriage of the Prince, and if they did engage in that practice, then surely the marriage of the Prince cannot explain that practice. The explanation might run now: 'The Prince married an American woman, and so the populace now began to put out American flags on the national holidays.' But the beginning of the practice is not what I am terming the beginning in a change. The beginning in the change is what the practices were *before* the causal event occurred which is purportedly the explanation of the practices.[1]

The general point to be made is that we can always refute an historical explanation (and indeed any causal explanation) of something x being in a certain condition G as a consequence of a proposed causal episode y, if we can demonstrate that x was G antecedently to the occurrence of y. This is particularly and obviously so when we propose explanations for the *origin* of something. Thus it is a plausible explanation of the origin of the practice of decorating evergreens at Christmastime to point out that in the late middle ages, in Alsatia, plays were presented the day before Christmas (Adam and Eve's Day) depicting the story of Paradise; that for scenic purposes apples were hung on trees, in this case evergreens, which would be the only trees bearing foliage at this season; and that given a natural human propensity towards embellishment, the decorations grew increasingly elaborate. One may refute this as an explanation of the *origin* of the Christmas tree—and refutations of this sort are very much the business of history—by discovering that evergreens were decorated at Christmastime earlier than this.

In view of these considerations, we can give a sense to what is sometimes offered in explanation of x being G, namely that x always has been G. This, of course, is simply to be understood as a way of saying that no special explanation is required for the fact that x is G *now*: a stranger may demand an explanation of why Jones is irascible this morning and be told that Jones is irascible *every* morning. Or one may ask why the monagasques put out flags on their national holiday and be told that they always have done so. But of course in either case the explanation, if we may term it such, may lead to a fresh demand for the origins of Jones's irascibility or the monagasque practice, and this, in turn, is a request for a cause, and hence implies a change.

The beginning, as specified in an explanadum, is then that state of x before x changed into its present condition. This fact permits the use of a special class of narrative sentences which *negatively* describe x as x was before the change, descriptions, again, which it would often be weird and pointless to suppose could have been given at that time. Thus an historian, writing the history of Alsace, might say, of a certain period, that the Alsatians had not yet developed their quaint custom of hanging apples on evergreen trees the day before Christmas. But, put into the present tense, we could hardly expect this sentence to appear, say, in the journal of a contemporary traveller in Alsatia. This would have the ludicrous effect as would a diplomatic message from Saxony to Paris in 1617 saying that the Thirty Years War had not yet begun—though nothing untoward is felt in an historian writing, of Europe at that period, that this was the eve of the Thirty Years War. Meanwhile, I can imagine a child suddenly acquiring the concept of the past by being told that there was a time, once, when people did not have Christmas trees. Etiological myths often begin with some such phrase as that.

Roughly, then, what we select as the beginning of a narrative is determined by the end, a claim borne out by the legitimacy of narrative descriptions of the beginning with reference to the end. One chief task in narration is to set the stage for the action which leads to the end, the description of which is the explanation of the change of which beginning and end are the termini. I have referred to *temporal wholes* in my discussion of narrative sentences, and suggested that it is characteristic of history that it organizes the past into temporal wholes. I realize that words like 'whole' are notoriously difficult to analyse, and that by 'whole' we are sometimes said to mean more than just a collection of parts. We mean a *unified* collection, and the chief difficulty, perhaps, has to do with the concept of unity. 'Unity', of course, is frequently a term of critical appraisal: we respectively commend or discommend a work of art in accordance with whether it has unity or not, whether or not its parts hang together. Doubtless we have differing criteria of unity, depending upon the genre of the work of art—poem, painting, musical composition—we are critically appraising, but I am here concerned only with the concept of unity as it applies to narration, and it seems to me we can make a beginning towards specifying a criterion of narrative unity by taking seriously the suggestion that a narrative and a deductive argument might constitute alternative *forms* of explanation. If this is so, then certain formal fallacies in deduction, certain deficiencies in an argument which prevent it from going through, might find their analogues in narratives. That is, we might find a number of conditions which, if an argument fails to satisfy them, render the *narrative* invalid. These would then be necessary conditions for a valid argument. By analogy we might constitute these as necessary conditions for a 'valid' narrative. I do not say we can elicit *all* the necessary conditions for narrative unity in this way, but we can get some. Moreover, by exploring the analogy between deduction and narration, we can begin to glimpse where the analogy gives way, and this will enable us, I hope, to determine what special and irreplaceable rôle narratives play in historical explanation.

I shall take the simplest case which will satisfy Hempel's criteria for an explanation, a *modus ponens* argument:

(1) $(x)Fx \supset Gx$,
(2) Fa,
(3) Ga.

where (3) is the explanandum (a sentence describing a singular occurrence), and where (1) and (2) jointly constitute the explanans, respectively as general law and initial condition. I shall assume that the explanans satisfies all of Hempel's criteria and that, moreover, (3) follows, by logic alone, from (1) and (2). I have already registered my dissatisfaction with this model, my main complaint being that Fa-Ga is a *change*, that this change is what we want an explanation for, that these changes are not always covered by general laws although the connection between these changes and some assigned *cause* for the change typically is covered by a general law, and so on. None the less, just with this simplified model we can make a few logico-aesthetic points.

(*A*) Suppose we were to replace (2) with Fb. This would be a violation of a rule in natural deduction, and the premises would no longer entail Ga. But similarly, suppose we were to replace (3) with Gb. This conclusion would no longer be entailed by (1) and (2). Logically, we want the same variable to be replaced by the same constants throughout. The narrative analogue might be spoken of as *unity of subject*. In the above argument, no constant can appear in the conclusion which does not antecedently appear in the premises. Narratively, 'the continuity or persistence of elements which a characteristically historical explanation emphasizes may be of a kind which serves to render the *explanandum*— when it is some human action or sequence of actions—intelligible or justifiable'.[1] There is an immensely difficult problem here in historical ontology, the problem, namely, of what *are* the elements which persist through a change. It is rather simple when we are concerned with the Duke of Buckingham's shift in attitude. But it is considerably more complex and metaphysically challenging when we are interested in such a change as, say, the break-up of feudalism, or the emergence of nationalism,

249

or, for that matter, the progressive embellishment of the Christmas tree. However this issue is to be decided, from a *formal* point of view a narrative requires a continuant subject.

(B) It is a commonplace in logical theory that no predicate may appear in the conclusion of a deductive argument which is not antecedently contained in the premises. Let us suppose that our conclusion satisfies condition (A) but contains an extra piece of information. For example, suppose it is a conjunctive assertion of two propositions about a—Ga and Ha. Clearly, (1) and (2) by themselves do not provide adequate evidence for the assertion of this conjunction, and the explanation would accordingly be incomplete. But an analogous point may be made about narratives, whether historical or fictional. Suppose that at the end of the play we know both that Macbeth is dead and that he is detested by Macduff, but that the play itself only accounts for the former fact. Since the author has (on our supposition) made a point of the fact that Macduff hates Macbeth, but has not shown us why he does so, we would feel a natural gap in the story and an artistic flaw in the play, the gap and the flaw being removed when episodes are introduced to account for Macduff's hostile attitude towards Macbeth: the fact that Macbeth caused the death of Macduff's wife and children. There may *in fact* be lots of things true of Macbeth at the end of the story for which no explanation has been furnished in the body of the play. From amongst all these only a certain few have been selected for narrative explanation. But of course this is so for any explanation, historical or otherwise. The point once again is that we do not explain events as such, but rather events under a certain description; it being as important in history as it is in science to choose a description. But once chosen, the event must be thoroughly explained relative to that description.

(C) Suppose we simply added a premiss (3a): Ea. The deduction would go through as before, but (3a) would make no contribution to the logical work. It would be superfluous and deductively inert. It violates a rule of deductive elegance, which requires that a valid deduction contain all and *only* those sentences required for the conclusion. Aesthetically, and with reference to a narrative analogue, this would represent a breach of artistic taste. It is a flaw in a narrative if it contains episodes which fail to contribute to the action. The scene with the drunken Porter in *Macbeth* is an instance of a narratively inert episode, and has indeed been criticized on just these grounds. This does not, of course, mean that its inclusion could not be otherwise justified. It may, for instance, provide some relief from the mood of intense horror created by the immediately prior murder scene. Historians too may introduce narratively inert information. But I am here interested only in the explanatory aspect of narratives.

On the basis of these analogies, then, we can, I think, state some of the necessary conditions for narrative unity. Thus, if N is a narrative, then N lacks unity unless (A) N is about the same subject, (B) N adequately explains the change in that subject which is covered by the explanandum, and (C) N contains only so much information as is required by (B) and no more. I do not say these are the only criteria for unity, and there may be other criteria for a satisfactory piece of historical writing which may even conflict with some of these, for instance, (C). But I do not wish unduly to complicate the issue under discussion by going further than this, for the rôle of narratives remains to be accounted for.

That there are these analogies between deductive arguments and narratives helps support my claim that a narrative is a form of explanation if a deductive argument is. But now I should like to show where it seems to me that the analogy gives way, and why, accordingly, narratives are not always reducible to deductive arguments.

I have so far represented a story, in a minimal way, as having some such form as this:

$$\text{(i)} \quad Fa,$$
$$\text{(ii)} \quad \gamma,$$
$$\text{(iii)} \quad Ga.$$

Here Fa and Ga together, and in that order, represent a change in a. This change may not be covered by a general law, but once reference is made to γ—a causal episode—then some general law is being appealed to, some general assumption is made, to the effect that γ-like things cause a-like things to change from F to G. Such a narrative I shall now term an *atomic narrative*, containing a beginning (i), and ending (iii), and a middle (ii). Graphically, I shall represent it as follows:

$$F\,G$$
$$|\cdot|$$

—where the strokes represent the termini of a change, and the dot represents the cause of the change.

But now it may happen that there are changes such that no single cause can serve to explain them. In this case, we may suppose that *a*—the subject of change—has gone through a *sequence* of changes, and that accordingly a sequence of causes must be assigned in order to explain the major change. Roughly in the way that a man, say, can only get from Westchester to Nairobi by taking (say) an underground train, a plane, a train, and a boat in that order, there being no single mode of transportation which will serve to carry him the whole distance. In cases of this sort, where no single cause can account for the change, but only a sequence of causes, each accounting for a successive change, I shall speak of a *molecular narrative*, to be represented graphically thus:

$$F\,G\,H\,I$$
$$/\cdot/\cdot/\cdot/$$

where the three succeeding changes are *F–G*, *G–H*, and *H–I*.

In a molecular narrative, each unit $/\cdot/$ is covered by a general law of the sort at least that I have characterized above, but there need not be any general law which covers the entire change. There may be a question as to why we require the notion of a molecular narrative, and cannot consider any such molecular narrative as simply an end-to-end series of atomic narratives. The answer to this is plain. It is because we are interested in the *larger change* (in the above representation) *F–I*, of which the intermediating changes are *parts*. But this serves, in turn, to answer what might seem an objection to our account, and indeed an objection to the entire enterprise of history. The objection is this: why, in order to account for *I*, do we need anything more than the last unit in the chain of atomic narratives. For is it not the case that the cause represented by the dot in

$$H\,I$$
$$/\cdot/$$

is the cause of *I*? In which case why go all the way back to

$$F$$
$$/?$$

The answer is this: the cause in question does serve to explain *I*, but the fact is that we are not specifically interested in *I* as such, but in the *change*

F–I, and for this change the cause cited is not sufficient. Here is one further instance of the sort of mistake which arises from thinking of the explanandum in an historical explanation as simply the description of an event. When, on the contrary, we see that the characteristic historical explanandum describes a change, and often, indeed, a *vast* change covering, perhaps, centuries, we can immediately see why we cannot reduce a narrative explanation to its final episode (or atomic narrative).

But now what of our analogy between narratives and deductive arguments? To begin with, it may be argued that, just as we must refer to several changes, and hence several causes, in order to account for a large 'molecular' change, so, in a deductive argument, we may need *several* distinct premisses in order to derive a conclusion, no single one of which by itself entails the conclusion. Thus, thinking of condition (B) above, we could not, for example, deduce Ha from just the two premisses Fa and $(x)(Fx \supset Gx)$. But by *adding* a further premiss, we can complete the argument validly. According to Hempel's model the added premiss must either be a general law or a statement of another required initial condition or both. Now suppose we add the general law $(x)(Gx \supset Hx)$. This would do the trick, but the fact is we can in such a case eliminate the *two* general laws in favour of another one, for since we can validly derive $p \supset r$ from $p \supset q$ and $q \supset r$, the two laws collapse into one—$(x)(Fx \supset Hx)$. But such an elimination cannot obviously be made in every valid narrative. Thus suppose we have

$$F\,G\,H$$
$$/\cdot/\cdot/$$

which presupposes two general laws, and refers to two distinct causes, say $k \cdot 1$ and $k \cdot 2$. The laws are as follows:

$$k \cdot 1 \quad F\text{-}G,$$
$$k \cdot 2 \quad G\text{-}H.$$

We may not be able to collapse these laws into a single larger one, and moreover we may not be able to find a single cause for the change *F–H*, so here the analogy would break down.

On the other hand, there is one further possibility. Suppose we eliminate our general law, and replace it with this *one*: $(x)(Fx \cdot Gx \cdot \supset \cdot Hx \cdot)$, and we add the further initial condition Ga. Here neither Fa nor Ga alone entail the conclusion, and the law requires their conjunction if its ante-cedent is to be satisfied. This would certainly be analogous to the case of a narrative which requires more than one cause to account for a large-scale change. Indeed, we might even have laws in which the required initial conditions are to be satisfied in sequence, for example, $(x)\,(Fx_{t-1}\,.\,Gx_{t-2}\,.\,\supset.\,Hx)$, where the subscripts indicate the order in which the initial conditions are to be satisfied. We might term such laws *historical* laws. Then, with the aid of historical laws, together with a speci-fication of temporally distinct initial conditions, we could indeed deduce our conclusion. Such laws would, in fact, enable us to make predictions, or better, *qualified* predictions. For since the two forms $(p\cdot q)\supset r$ and $p\supset(q\cdot r)$ are demonstrably equivalent, it follows that if we have an *historical law* of the form

$$\left(C^0_{t-0}\,.\,C^I_{t-1}\,\ldots\,.\,C^n_{t-n}\right)\supset E$$

and, if C^0 occurs at t–0, we can predict that E will take place *if* C^I ... C^n take place in the required temporal order. Roughly in the way that we might say (predict) that a rocket will cover a certain distance providing three serial explosions occur, and then, supposing the first *does* occur, go on to predict that the rocket will cover the required distance if the two *remaining* explosions take place.

There may *be* historical laws. There may even be historical laws in history, for all one knows. But should they be discovered they would not add any further support to determinism than would the existence of non-historical laws. Nor would they in any way entitle us to conclude that there is historical inevitability any more than the existence of general laws of a non-historical sort entitle us to conclude that there is inevitability in nature. So the discovery of historical laws would in no degree support the prophetic pretensions of substantive philosophers of history.

Meanwhile, I think, I may justifiably claim that whether we can transform a molecular narrative into a deductive argument is contingent upon the question whether there exist historical laws, and moreover there would remain, even if *some* historical laws should be discovered, the question whether for *each* molecular narrative a general historical law might be found.

Be this as it may, the fact remains that, in history at least, few, if any historical laws are known, but this in no way diminishes or jeopardizes the explanatory force of narratives. If anything, it jeopardizes a philosophical programme which is committed to the view that every explanation requires, as a necessary condition, that it be capable of deductive formulation. I have conceded that this may very well be a *sufficient* condition, but not a necessary one if molecular narratives be accepted as explanatory. This does not again entail that narratives can be constructed without the use of general laws, but only that no general law need be found to cover the *entire change* covered and explained by a narrative.

The rôle of narratives in history should now be clear. They are used to explain changes, and, most characteristically, large-scale changes taking place, sometimes, over periods of time vast in relationship to single human lives. It is the job of history to reveal to us these changes, to organize the past into temporal wholes, and to explain these changes at the same time as they tell what happened—albeit with the aid of the sort of temporal perspective linguistically reflected in narrative sentences. The *skeleton* of a narrative has this form:

$$/./././. \ldots ./$$

but the skeleton may be fleshed out with extra descriptions, anecdotes, moral judgements, and the like. But these, I am suggesting, are, philosophically at least, of secondary interest.

One final word. Even supposing we had really extraordinary historical laws, involving vastly many variables and covering immense stretches of time, there would still be no reason to suppose that the connection between these laws and the temporal wholes which instantiate them would be any less loose than the connection between the laws I have been discussing and their instances. So not merely would such predictions as we might make by means of these laws be conditional, but they would also be general. They would tell us at best what will happen only under certain highly *general* descriptions providing that certain initial conditions—again under highly general descriptions—sequentially hold. So once again the prophetic aspirations of substantive philosophers of history would be subverted, and once again there would be the problem of writing the history of events before the events happened, a task which the existence of historical laws would not enable us to discharge. Our knowledge of the future would remain abstract in contrast with our knowledge of the past. And the task of history itself would still be to tell the story of what precisely happened, even if the story should fall under a general historical law as an instance, and even if the law should be known. History alone would be able to exhibit the amazing variety of temporal wholes which none the less all fall under a single historical law. Our fascination with the *details* of the past would, if anything, increase. One does not find sonnets less interesting or beautiful upon being told that all sonnets have an invariant form. If anything our admiration for poetic creativity increases upon learning that so many distinctly individual and dissimilar works should all have been produced in conformity with the most rigid and invariant set of rules!

XII

METHODOLOGICAL INDIVIDUALISM
AND METHODOLOGICAL SOCIALISM

I have sought to make out a case against substantive philosophy of history by emphasizing certain logical features of what one might call the language of time. And in the course of this I have tried to explicate our concept of history, suggesting that it is the illicit extension of modes of description which are essentially historical beyond the domain in which they have application which defines the aspiration of substantive philosophy of history. I have sought, then, to draw a borderline which we are tempted to cross but cannot. The analyses of history and of substantive philosophy of history are interdependent, roughly in the manner in which the Transcendental Analytical and the Transcendental Dialectic are in Kant's *Critique of Pure Reason*: for the Dialectic exhibits the unhappy destiny which attends Reason when it seeks to extend those forms of understanding specified in the Analytic beyond the domain where they have application, the domain, namely, of experience. It is in the spirit of critical philosophy that I should wish to have my argument understood.

To this case, however inadequately it may have been stated, I have nothing further to add. But before ending I should like to dismiss another kind of charge, which is sometimes made against substantive philosophy of history, namely that it holds the view that history is not something which men make, but that the moving agents of history are certain superhuman or superorganic entities. I do not wish to say that it is innocent of this charge, but rather that it is no philosophical crime to be guilty of it; and if philosophical historians subscribe to such entities, they may or may not be wrong, but their wrongness will be factual and empirical, and not conceptual and philosophical. To show this, however, requires some detailed conceptual analysis, for the issues here are exceedingly tangled, and the discussion is composed of topics in ontology, meaning, methodology, and language which resemble one another almost too much for us to expect enthusiasts to take the time to sort them out. I shall endeavour to do this now, hoping, in transit, to contribute somewhat further to the correct analysis of historical sentences.

By *historical sentence* I shall mean: a sentence which states some fact about the past. Historical writings consist chiefly of historical sentences, and are further distinguished by the fact that a considerable number of the historical sentences which compose them employ, as grammatical subjects, proper names (e.g. 'Frederick V') or definite descriptions (e.g. 'The Elector Palatine in 1618') of individual human beings who actually existed. Neither of these linguistic features carries us very far towards an adequate characterization of historical writings. Historical sentences do not uniquely occur in historical writings, and while it is perhaps (logically) inconceivable that there should be an historical writing which contained *no* historical sentences, there is little difficulty in conceiving an historical writing in which *none* of the historical sentences which compose it employs expressions which refer directly to individual human beings who actually existed. Indeed, an historical writing *all* the sentences of which employed such expressions would be far more difficult to imagine.

Individual human beings are not the only individuals directly referred to by the subjects of historical sentences. There are, in addition, what I shall term *social individuals*, individuals which we may provisionally characterize as containing individual human beings amongst their parts. Examples of social individuals might be social classes (The German bourgoisie in 1618), national groups (the Bavarians), religious organizations (the Protestant Church), large-scale events (The Thirty Years War), large-scale social movements (the Counter-Reformation), and so on. Social individuals are not the only kind of individuals, other than individual human beings, which may be referred to by the subjects of historical sentences, so these two kinds of individual-referring historical sentences do not make up the entire class of historical sentences. Nevertheless, it is with just these two kinds of sentence that I shall be concerned here.

It will, I think, be universally admitted that, from the point of view of ready communication, neither kind of sentence could easily be eliminated from the language of the historian. It is difficult to see how, for example, an historian could convey, by means only of sentences which referred directly to individual human beings, the information so neatly and intelligibly communicated by 'The Thirty Years War began in 1618'. Never the less, ease of communication and considerations of narrative economy notwithstanding, certain philosophers and historians have expressed a certain mistrust with regard to this latter kind of sentence— a mistrust the source of which lies in mistrust with regard to social individuals as such. These thinkers exhibit a pardonable reluctance to concede that the social world is made up of individual human beings *and* other, super-human individuals which, though they may be said to contain human beings amongst their parts, none the less are not wholly to be identified with these parts, and which enjoy, so to speak, a life of their own. In some sense or other, they appear to be saying that the social world is made up of individual human beings alone, and that there is nothing which both contains individual human beings amongst its parts and is itself an ultimate occupant of the social world. So at first glance what seems to be in issue is a question in ontology. Thus Mr J. W. N. Watkins, a philosopher, writes:

The ultimate constituents of the social world are individual people who act more or less appropriately in the light of their understanding of their situation. Every complex social situation, institution, or event is the result of a particular configuration of individuals, their dispositions, beliefs, and physical resources and environments.[1]

And Professor H.-I. Marrou, an historian who otherwise never wearies of cautioning us that there are more things on earth and in heaven than are dreamed of in our philosophy, writes:

Ce qui 'a réelment existé', ce n'est ni le fait de civilisation, ni le système ou le supersystème, mais l'être humain dont l'individualité est le seul véritable organisme authentiquement fourni par l'expérience.[2]

The use of the word 'civilisation' here is almost certainly polemical, to be taken in antagonistic reference to Professor Toynbee's well-known thesis that civilizations do indeed have a life of their own, and that they, moreover, constitute the least units of historical study. But in fact he is rejecting all allegedly superhuman entities in history:

A lire certains travaux contemporains, on a l'impression que les acteurs de l'histoire ne sont plus *des* hommes, mais des entitiés, la Cité antique, la féodalité, la bourgeoisie capitaliste, le proletariat révolutionnaire. Il y a la un excès.[1]

And Watkins impugns in general the view that 'some superhuman agents or factors are supposed to be at work in history'—for example 'the alleged long-term cyclical wave in economic life which is supposed to be self-propelling, uncontrollable and inexplicable in terms of human activity'—and insists that 'human beings are . . . the sole moving agents in history'.[2]

These writers exhibit in these passages such an essential unanimity of attitude, that it must seem sheer pedantry to distinguish between their positions. Nevertheless, it seems to me that they are in fact maintaining distinct, and possibly even independent views. Professor Marrou's views seem quite unequivocally to be a thesis in ontology. He is insisting that, in the social world, *only* human individuals are real, and superhuman, or social individuals, are not. His criterion for applying 'x is real' is patently an epistemological one: 'x is real' if, and only if, we experience x. Mr Watkins, by contrast, *may or may not* be defending an ontological thesis. He has said only that human beings are the *ultimate* constituents of the social world, and the context quite clearly implies that he means, by 'ultimate', that human beings are the only *moving agents* in the social world. It does not follow from the fact that human individuals are the sole moving agents in the social world, that they are the only members of the social world, but only that, whatever else may be a member of the social world is not a moving agent. So Watkins, whether in addition he would also subscribe to Marrou's ontological thesis, is actually concerned with a thesis about *explanation*. What he seems to be saying is this: if indeed there are such things as social individuals, their behaviour is to be explained, ultimately, with reference only to the behaviour of individual human beings, and *not* with reference to the behaviour of other social individuals. This is because (he would claim) human beings alone are *causal* agents in history.

There are, as we shall see, further differences it will be useful for us to emphasize. But, because each of their positions bears a similarity to a third possible thesis, the thesis, namely, that sentences which ostensibly refer to social individuals are 'reducible' to sentences which only refer to individual human beings, it is important to recognize that neither of them is advocating this sort of reductionist programme. The strategy of such philosophical reductionism in current or recent philosophical discussion is this: if we have a set $\{S\}$ of sentences which employ terms of kind T ostensibly referring to objects of kind O, and another set $\{s\}$ of sentences employing terms of kind t ostensibly referring to objects of kind o, then, if we can replace every context ... T ... in $\{S\}$ with one or more contexts ... t ... in $\{s\}$, we will only need to admit objects of kind o in our ontology. If such a programme were in fact to succeed, of course it would not *follow* that there are no objects of kind O, but only that we do not require to *suppose* that there are such objects. Neither author is suggesting that sentences ostensibly referring to social individuals is eliminable in this or *in any other sense* from the language of historians. Indeed, I think, even if Professor Marrou were adamant in rejecting social individuals, he might still feel that sentences of this sort have an indispensable rôle to play in historical writings. And for him, at least, this rôle does not consist in expressing some fact about social individuals but perhaps, instead, in expressing, in some fashion or other, facts about individual human beings which perhaps could not as readily be expressed by means of sentences which directly are about individual human beings, if indeed those facts can be expressed by means of those sentences at all.

At this point I deem it advisable to introduce a concrete example in which it seems to me that an historian is expressing some fact about a social individual, however we are finally to analyse the concept of a social individual. In this example, the historian is seeking to explain a change, in which, to be sure, masses of individual human beings were in one manner or other involved, but a change, never the less, which they were very likely not aware of. This change took place 'insensibly', and was completed after roughly seventeen years.[1] It is a change described as follows: the loss of 'whatever spiritual meaning the [Thirty Years War] had'. It will prove particularly instructive to note in what way reference to individual human beings is made. Miss Wedgewood describes the background of the change in this way:

While increasing pre-occupation with natural science had opened up a new philosophy to the educated world, the tragic results of applied religion had discredited the Churches as the directors of the state. It was not that faith had grown less among the masses; even among the educated and the speculative, it maintained a rigid hold; but it had grown more personal, had become essentially a matter between the individual and his creator . . . .

Inevitably, the spiritual force went out of public life, while religion ran to seed amid private conjecture, and priests and pastors, abandoned by the state, fought a losing battle against philosophy and science. While Germany suffered in sterility, the new dawn rose over Europe, irradiating from Italy over France, England, and the North. Descartes and Hobbes were already writing, the discoveries of Galileo, Kepler, and Harvey had taken their places as part of the accepted stock of common knowledge. Everywhere, lip-service to reason replaced the blind impulses of the spirit.

Essentially, it was only lip service. The small group of educated men who appreciated the value of the new learning disseminated little save the shadow of their knowledge. A new emotional urge had to be found to fill the place of spiritual conviction; national feeling welled up to fill the gap.

The absolutist and representative principles were losing the support of religion; they gained that of nationalism. That is the key to the development of the war in the latter period. The terms Protestant and Catholic gradually lose their vigour, the terms German, Frenchman, and Swede assume a gathering menace. The struggle of the Hapsburg dynasty and its opponents ceased to be the conflict of two religions and became the struggle of nations for a balance of power . . . .[1]

Briefly, and with extraordinarily sure touches, Miss Wedgewood is here describing and explaining the way in which the Thirty Years War changed from a religious to a political conflict, and when she makes reference to individuals, it is mainly in order to *illustrate* this change, or to provide evidence that a change has in fact taken place. It is, as it were, taking soundings in the flow of history. Here are some examples:

The aging Emperor, the Electors, of Saxony, Brandenberg, and Bavaria, the Swedish Chancellor, Richelieu . . . still held their course. But all around them had grown up a new generation of soldiers and statesmen. War-bred, they carried the mark of their training in a caution, a cynicism, and a contempt for spiritual ideals foreign to their fathers.[1]

And again:

Frederick of Bohemia had lost his crown because he had offended subjects in order to obey his Calvinist chaplain; his son, Prince Rupert, Calvinist in religion and morality, fought in England against Presbyterians and Independants, because his religion was for him, as for most of his generation, nobody's business but his own.[2]

These various individuals are selected for special mention, and certain facts about them singled out and contrasted with one another, not because of any intrinsic interest they may have—for indeed they may have no intrinsic interest—but because of their *historical* significance: they make clear to us that a great change in attitude and behaviour of individuals in roughly the same social positions has taken place. Consider one further example. The battle-cry shouted by soldiers at White Hill was 'Sancta Maria!' The battle-cry shouted, later, at Nordlingen, was 'Viva España'. Those who might have witnessed these two battles would almost certainly not have seen the significance of these shouts. For the significance lies in the contrast between them, a contrast which is significant to an historian who sees in them signs that 'insensibly and rapidly, the Cross gave way to the flag'.

I think it reasonably certain that very few of these changes were *intended* by anyone. Men followed their own purposes, acted in the light of their views of their situations, were not alive to the 'significance' of what they were doing. Moreover, the changes here described may not have been reproduced within the biography of any *single* individual who lived through the change: to modify slightly an example as old as Leibniz, we may change the colour of a dish of blue powder by the device of adding yellow powder, so that the content of the dish changes from blue to green without any *single* particle having to change colour. Even if certain individuals did in fact change, we still could not, from this alone, deduce the scope of the changes which incontestably took place. Even if we had complete biographies of every individual human being alive during this period, we would still have to make careful comparisons and contrasts to infer that changes of this sort had happened. To put the matter baldly, the change took place, not in individuals, but in *society*.

Now it seems to me that what took place in the period covered by this account is a fairly clear example of what Watkins terms 'organic-like social behaviour', which is to say that

Members of some social system (that is a collection of individuals whose activities disturb and influence one another) mutually adjust themselves to situations created by others in a way which, without direction from above, conduces to the equilibrium or preservation or development of the system.[1]

Watkins further says that

Such far-flung organic-like behaviour involving people widely separated in space and largely ignorant of one another, cannot be simply observed. It can only be theoretically reconstructed—by deducing the distant social consequences of the typical responses of a large number of people to certain repetitive situations.

Miss Wedgewood's account is an instance of this for a variety of reasons, but for the present I wish to draw attention to the fact that the social changes she has drawn our attention to are *not*, as such, observable. All that can be observed, indeed—waiving for now questions concerning problems of verifying, through observation, statements about the past—is the behaviour of individual human beings. But this behaviour is never the less to be appreciated and understood with reference to a social system in course of modification, and is to be taken as *evidence* for the fact that the system itself is being modified. It would only be through accepting the most thoroughgoing verificationism that we could say that her account was *about* the individuals, statements about whose behaviour provide, perhaps, the only evidence in the nature of the case available for the verification of statements about the changing social system.

With this in mind, we may now recognize those further differences, earlier alluded to, between the positions of Watkins and Marrou. The fact that we cannot observe social systems as such, but only the behaviour of a set of individuals, must entail, by Marrou's criterion, that social systems as such are not 'real'. Never the less, he is prepared to concede that the concept of a social system may have a theoretical use, in the sense that we might require such a concept in order to explain the behaviour of individual human beings. Consistent with this, I suppose, he would insist that electro-magnetic fields are not 'real' either, and that the use of field theories in no way commits us to supposing fields part of the physical world. Fields, like social systems, would be 'abstractions'. Thus:

Même s'il apparaît a l'examen de toutes données documentaires, que tel phénomène historique s'explique par l'un de ces abstraits socio-culturels, l'historien devra toujours se garder d'oublier et de laisser oublier, que ce n'est là qu'un construction de l'esprit, inévitable, sans doute (comme étant le seul moyen de saisir la complexité du réel) et, dans les limites d'emploi, légitime—mais tout de même une abstraction, un produit dérivé, et non pas le réel lui-même, ni, comme on finit toujours par le croire, du surréal![1]

Watkins, by contrast, seems quite prepared to allow that the social world contains, in addition to individual human beings (who may never the less be its 'ultimate components'), social systems, the behaviour of which is 'organic-like'. In fact he is concerned precisely with the explanation of just such things as these, but quite oppositely to Professor Marrou, he does not believe—and this is his main contention—that we in any ultimate way require reference to facts about such social systems in order to explain other facts about them. Theories which do make that sort of reference are what he terms 'half-way theories', and these are contrasted with 'rock-bottom explanations':

We shall not have arrived at rock-bottom explanations of such large-scale social phenomena until we have deduced an account of them from statements about the dispositions, beliefs, resources, and inter-relations of individuals.[2]

And the plain suggestion here is that Watkins does not believe that, in speaking of such large-scale social phenomena, we are merely speaking of individual human beings and their beliefs, dispositions, resources, and inter-relations, for then the distinction between 'half-way' and 'rock-bottom' explanations collapses.

I shall now begin to concern myself in a critical way with Watkins's position, which he has termed *Methodological Individualism*,[3] for it seems to me that Professor Marrou's views can best be appreciated and assessed in the light of such a discussion.

Let me begin by emphasizing that there are a number of closely related but never the less distinguishable thesis which must be carefully kept apart from Methodological Individualism. Here are some things it is *not*.

(1) It is *not* a theory of meaning. It does not hold that every statement about social phenomena is 'really' or 'ultimately' a statement about individual human beings. Nor does it propose to demonstrate that every *predicate*, nominally true of social individuals, may be explicitly defined by means of predicates which range over individual human beings. Hence it is not an *analytical* theory, in accordance with which *sentences* about social individuals are held to be, at least in principle, translatable, without loss of meaning, into sentences wholly about individual human beings. On the other hand, it *does* require that there be *some* kind of relation between these two classes of sentence. For instance, it may very well be that only through verifying, by observation, certain sentences about individuals, shall we ever be able to confirm a sentence about social systems. But this does not mean that sentences about social individuals are really to be understood as about that, the observation of which will confirm them. The Methodological Individualist is not neccessarily committed to the Verifiability Criterion of meaning. Indeed, as we shall see, it is important that these two classes of sentences have, and retain, their distinctive sorts of meanings. In general then, it would be irrelevent to demonstrate that sentences about social individuals (or social phenomena however understood) are 'irreducible'.

(2) If it is not an analytical theory, neither is it a constructionist theory. It does not subscribe to Russell's celebrated dictum that inferred entities should always, in the interests of parsimony, be replaced with logical constructions. It does *not* hold that societies are logical constructions out of individuals in the way in which Russell used to say that stars and tables are logical constructions out of sense-data. The Methodological Individualist is not after a *Logische Aufbau der Gemeinsschafftswelt*. Such a programme, like its analytical counterpart, might be philosophically interesting and intellectually challenging and possibly even important, but the viability of Methodological Individualism does not depend upon *its* viability, nor would the shipwreck of such a programme be of more than external interest to the Methodological Individualist. When Watkins speaks of *constructions*, he has in

mind the construction of a theory of a scientific, and not a metaphysical sort, and whose purpose is not to eliminate, but to account for social systems. Such a theory indeed has sentences about individual human beings in its *base*, but we must distinguish between the base of a theory and the rest of the theory; quite clearly the concept of a *base* loses all meaning if there is not something else, distinguishable from it, for which it *is* the base: a building cannot be *all* foundation.

(3) Methodological Individualism is not an ontological theory, in accordance with which only human beings are real in the social world. Someone may wish to say that societies, or social individuals, depend for their existence upon individual human beings, and that if there were no individual human beings, there would be no social individuals either. But plainly, if something exists contingently, it still exists. The Methodological Individualist is not motivated by a metaphysical monism, and the controversy between him and his opponents is not analogous to the controversy between those who hold that images *are* simply brain-states, and those who deny they are brain-states. Indeed, he is almost militantly dualist. His position *is* analogous to the epiphenomenalist who, holding that images (and mental events generally) are distinct from brain processes (and physical events generally), still insists that the former are causally connected with the latter in a unilateral way, and can only be *explained* with reference to the latter.

(4) Accordingly, the Methodological Individualist does not deny in advance that there are, or may be found, general law-like sentences which relate various properties of social systems. Nor does he maintain that such laws, should they be found, would really only be laws describing the behaviour of individual human beings. He might, indeed, as I have already suggested, insist that such laws can be confirmed only through the verification of sentences about human beings. But this would in no way entail that the laws in question are not really laws which describe the gross behaviour of social individuals.

So we see that Methodological Individualism has nothing whatsoever to do with a number of interesting and exciting positions it might be thought to resemble. Very briefly, it appears to hold (*a*) that sentences about social individuals are logically independent of sentences about individual human beings, (*b*) that social individuals are ontologically distinct from individual human beings, (*c*) that social individuals are causally dependent upon the behaviour of individual human beings and not the other way about, (*d*) that explanations of the behaviour of social individuals are always to be rejected as ultimate unless these explanations are framed exclusively in terms of the behaviour of individual human beings, and (*e*) the explanation of the behaviour of individual human behaviour must never be in terms of the behaviour of social individuals; (*a*) is a thesis about meaning, (*b*) and (*c*) are theses about the world, and (*d*) and (*e*) are theses about the ideal form of a social science.

Now the natural methodological position which is opposed to Methodological Individualism I shall term *Methodological Socialism*. And once again, there may be positions, analogous to (1)–(4), which resemble but are never the less distinct from Methodological Socialism. Thus someone might hold that sentences about individual human beings can be translated into sentences about social individuals, or that social individuals are real and that human beings are not, and so on; Hegel seems to have subscribed to such views as these. But to avoid any pointless diversions from our main concern, I shall characterize Methodological Socialism as follows: simply replace every occurrence of 'individual human beings' in the theses (*a*) through (*e*) in the preceding paragraph, with the expression 'social individuals', and every occurrence of 'social individuals' with 'individual human beings'. The resulting sentences (*a*)–(*e*) will now characterize Methodological Socialism. Notice that (*a*) and (*b*) are left unaffected by the substitutions. The controversy between the two positions turns, accordingly, on (*c*), (*d*), and (*e*), and in whatever sense the Individualist will say that individual human beings are *ultimate* in the social world, the Socialist will say that *social individuals* are ultimate in the social world.

Not surprisingly, the most conspicuous example of a theory which satisfies this specification of Methodological Socialism is Marxism, and not surprisingly, Marxism has been one of the main targets of the Methodological Individualists, e.g. Watkins and Popper. One need but think of how subscribers to either of our methodological positions would *explain* this fact, in order to produce specific explanations which very nicely fit the characterizations I have given of these two positions. Now that in Marxism which primarily illustrates Methodological Socialism is that part of Marx's theory known as *historical materialism*. Marx believed (and believed himself to have shown) that there is a one-way interaction between social processes and at least some psychological processes, so that what we think, and how we act, are to be explained by reference to our relations *vis-à-vis* the prevailing system of production; and whatever it is that causes changes in the system of production, it is *not* something which is brought about by individual human action. Crudely put, we explain some facts about systems of production with reference to other facts about systems of production; and we explain some facts about individual human beings with reference to some facts about systems of production; but we never explain any facts about systems of production with reference to any facts about individual human behaviour; and finally, we never explain any facts about individual human behaviour with reference to other facts about individual human behaviour.

Letting S stand for some fact about a social individual, and P some fact about individual human psychology, we may represent this position schematically as follows:

$$P_1 \qquad P_2 \qquad P_3$$
$$\uparrow \qquad \uparrow \qquad \uparrow \qquad\qquad\qquad (A)$$
$$S_1 \longrightarrow S_2 \longrightarrow S_3 \longrightarrow$$

where the arrows indicate the directions of causality, and where the *absence* of an arrow between any two points signifies the absence of a causal tie.

A theory which, on the other hand, satisfies the conditions of Methodological Individualism, would look schematically as follows (using just the same conventions as above):

$$P_1 \longrightarrow P_2 \longrightarrow P_3 \longrightarrow$$
$$\downarrow \qquad \downarrow \qquad \downarrow \qquad\qquad\qquad (B)$$
$$S_1 \qquad S_2 \qquad S_3$$

Meanwhile, Methodological Socialism will tolerate provisional, or 'half-way' theories of this form:

$$P_1 \ldots\ldots \quad P_2 \ldots\ldots \quad P_3 \ldots\ldots \qquad\qquad (C)$$

where the broken lines indicate only apparently causal ligatures. And *mutandis mutandum*, Methodological Individualism will allow provisional or half-way theories of this form:

$$S_1 \ldots \ldots \quad S_2 \ldots \ldots \quad S_3 \ldots \ldots \qquad \text{(D)}$$

It is important to stress that Marxism is only an *example* of a theory which could satisfy Methodological Socialism, and that (B) is only an *example* of a theory which would satisfy Methodological Individualism. So the latter is not in fact committed in any sense whatsoever to the *independent* thesis that we can explain facts about social individuals with reference to *psychological* facts about individual human beings. That is to say, Psychologism is only a particular example of a theory which satisfies the criteria of Methodological Individualism.[1] But surely no one, not even the most militant Individualist, would say that for this reason alone it is *correct*. At best he would say that such a theory is *acceptable*.

The issue now before us is whether there is any good reason for choosing between Methodological Individualism and Methodological Socialism. I shall examine at this point an extremely interesting discussion, by Professor Maurice Mandelbaum,[2] which, if I understand it properly, is intended to furnish just such good reasons for *rejecting* Methodological Individualism, at least as a general methodological programme. Professor Mandelbaum has argued for the existence and autonomy of what he terms *societal facts* which, he maintains, are 'as ultimate as psychological facts' and 'cannot be reduced without remainder to concepts which refer to the thoughts and actions of specific individuals'.[3] Societal facts 'are facts concerning the forms of organization present in society'.[4] About the relations between psychological and societal facts, Mandelbaum says a number of things. I shall set down those that strike me as his four main theses.

(i)   Sentences about societal facts cannot be translated into sentences about psychological facts without leaving a societal remainder.[5]

(ii)   Sentences about societal facts must be ('it is ... necessary to ...') partially translated into sentences about individual facts, for 'unless we do so we have no means of verifying any statements we may make concerning societal facts'.[1]

(iii)   Societal facts *may* depend for their existence 'upon the existence of human beings who possess certain capacities for thought and for action'. But this does not entail that the former set of facts is *identical* with the set of facts upon whose existence it is contingent.[2]

(iv)   The existence of societal facts does not entail that there *are* no individual facts, or that individuals are not 'real'. Rather, there are these two distinct sets of facts which may be said 'to interact'. Thus: 'There are societal facts which exercise external constraints over individuals, no less than there are facts concerning individual volition which often come into conflict with these restraints.'[3]

Now it should be plain that very little of this is incompatible with Methodological Individualism as I have characterized it, but at best it is incompatible with some of the things which might be thought to *resemble* Methodological Individualism. The sole point of conflict comes in thesis (iv) where, since quite clearly *two* way interactionism is intended, there is an incompatability with (*c*), and with the thesis which animates Mandelbaum's argument, namely:

(v)   In understanding or explaining an individual's actions, we must often refer to societal facts, i.e. facts concerning the organization of the society in which he lives.[4]

Let us now take a close look at Mandelbaum's arguments for (v). It consists of two parts, an informal (or Wittgensteinian) part, and a formal (non-Wittgensteinian) part. The informal part is this: one could not teach someone, who is strange to our culture, what a member of our culture is doing when he presents a withdrawal-slip to a bank-clerk, without, in process of doing so, teaching him something about the way in which withdrawal-slips function in that system of operations which make up our banking system. But since reference to a banking system is *ipso facto* reference to a societal fact, we cannot explain to someone, who is strange to our culture, what such a man is doing, without bringing into our explanation reference to a societal fact. Now notice how the word *explain* is being used here. It is being used in *this sense*: explaining *to* someone what that person does not understand. It is the sense in which a teacher explains a lesson, or the meaning of a word. In Mandelbaum's example, the teacher is making explicit what members of our own culture take for granted, and apply, perhaps, in a categorial kind of way when they are engaged in banking operations.

The *formal* part of the argument is essentially the translatability thesis (i). What Mandelbaum says here is this. Suppose we have a language L_s in which only terms designating societal facts appear, e.g. 'institution', 'mores', 'ideologies', 'status', 'class'; and suppose there is another language L_i which contains terms only for describing the thoughts, actions, and capabilities of individual human beings. Now Mandelbaum insists that we cannot translate sentences from L_s to L_i 'without remainder'. In fact, Mandelbaum is not quite sure of this point. He even suggests that it may be theoretically possible to effect this translation after all, only it would be immensely difficult in practice, and of no scientific interest.[1] But the point, he admits, is not really a matter of how societal concepts are to be analysed, but how individual actions are to be *explained*, so this brings us back to the informal part of the argument.

The two-language thesis I regard as a confusion. It merits a few words on that account, despite Mandelbaum's concession that, since no matter of principle may be involved, but only a matter of practice, no philosophical importance attaches to it. The fact is, that if we could not make the translation, this would be more damaging to his position than if we could. For we use L_s to talk about societal facts, and L_i to talk about individuals, and what he wants is a way of talking about individuals which makes reference to societal facts. So what he requires, I think, is this. We have some terms which refer to individual human thought and action which presuppose something about societal facts, and some terms which refer to individual human thought and action which do *not* involve that sort of presupposition. I do not know if there is in fact the latter sort of term, but let us suppose there is. Now any term, the correct application of which to an individual human being presupposes some fact about the organization of society, I shall call an S-predicate, and any term the correct application of which to an individual does *not* involve such a presupposition, I shall call an I-predicate. A sentence which employs an S-predicate will be called an S-sentence. Any sentence which uses *I*-predicates *alone* (i.e. as non-logical terms) I shall call an *I*-sentence. Thus:

(s-1) The bank-teller certifies the withdrawal-slip,
(s-2) The man makes marks on the piece of paper.

are respectively instances of an *S*- and an *I*-sentence. But

(s-3) The man certifies the withdrawal-slip

is also an *S*-sentence, since it does not use an *I*-predicate *alone*, but also uses an *S*-predicate.

We may use all three of these sentences to describe one and the same event. We may further say that (s-1) and (s-3) do, but that (s-2) does *not*, presuppose some fact about the organization of our society. A Trobriand Islander could be taught to understand (s-2) without being taught, in the process, any such facts. We may indeed say that *I*-predicates are trans-cultural. Any term, the meaning of which can be taught without bringing in some fact about the organizational peculiarity of a given society, naturally belongs to the *I*-vocabulary. We may now reconstruct Mandelbaum's thesis as follows: an *S*-sentence cannot be translated without remainder into an *I*-sentence or a set of *I*-sentences. The question now is how 'without remainder' is to be interpreted. I shall interpret it in a way which will, I hope, not merely be congenial to Mandelbaum, but which will demonstrate my revised version of his thesis.

Consider the following lists of terms, the ones in the right-hand column being *I*-terms which correspond to the *S*-terms in the left-hand column. The terms in either column refer to an individual human being, an action, and a material object respectively:

bank-teller	man
certifies	makes marks on
withdrawal-slip	piece of paper

Between the correspondent terms in either column, there are no entailment relations. We cannot deduce that someone is a man from only the information that someone is a bank-teller, nor vice versa. Not every bank-teller is a man nor every man a bank-teller; not every piece of paper is a withdrawal-slip, and even if every withdrawal-slip were *per accidens* a piece of paper, no logical difficulty would face a bank which resolved to introduce plastic withdrawal-slips; and finally, one may certify a withdrawal-slip without making marks on anything (the bank-teller could just say 'O.K.'), and it is plainly true that not every instance of making marks on something is a case of certifying something (not even when the marks are made on a withdrawal slip: the teller might write 'No good.'). Hence it is quite feasible that if '. . . man . . .' is offered as a translation of '. . . bank-teller . . .', the former could be false, while the latter is true; and since the weakest condition we can place upon translations is that they be *equivalent*, the former would not be a translation of the latter.

But now there *may* be societies in which such entailments hold. Thus, it may have been the case in a certain period in English history, that only males could be bank-tellers. But this would be only in virtue of some organizational feature, hence societal fact, having to do with a temporary peculiarity of British society. So the translation could go through only through presupposing *this* sentence: 'All bank-tellers in this society are male.' And *this* would be an *S*-sentence.

By 'without remainder' I shall therefore mean: you cannot translate an *S*-sentence into an *I*-sentence without presupposing another *S*-sentence. This is generally true. Hence anyone who proposed to *eliminate*, by translation, *S*-predicates in favour of *I*-predicates, could only do so on the basis of allowing exactly the sort of sentences he proposed to eliminate. Hence any such programme is inherently self-defeating, and the programme is demonstrably impossible.

What, however, are we more precisely to understand by these societal facts, implicit reference to which is presupposed in any such translation or, for that matter, in instructing cultural outsiders? I suggest that they are rules, norms, and conventions, as, 'Only males can be bank-tellers.' Indeed, it seems to me that this is the most natural way in which we may interpret Mandelbaum's thesis (iv), for we naturally speak of people being constrained to act in accordance with this rule or that, and of people chaffing under these rules and wishing to have them changed. Thus a woman whose great ambition in life is to become a bank-teller in a major bank might find herself frustrated by such a rule, and this might, with some latitude, be regarded as an instance of the interaction between an individual and a social fact.

But understood in this way, it is hard to see how either (iv) or (v) is any longer incompatible with Methodological Individualism. The Methodological Individualist is clearly not denying that there are rules, that people act in conformity with them, change them, are frustrated by them, and so on. As for (v) it has turned out to be very little more than a thesis about *meaning*, and the Individualist may very well agree with Mandelbaum that, when called upon to explain a given piece of behaviour to a stranger, in exactly the sense of 'explain' which applies in such contexts, we may have to tell him some of the rules, norms, and conventions. It may further be agreed that, in the categorial sense mentioned earlier, we understand actions, under *certain* descriptions of them, with reference to some rules, norms, and conventions. But this is *not* the sense of 'explain' which interests the Methodological Individualist. Rules, he may go on to point out, are broken all the time, and when a rule is broken, it is not by that fact alone to be considered *abrogated*. Only rules which *hold* can be broken. But he will add that his interest lies in *laws*, in the precise sense in which we speak of scientific laws. When such a law is broken, this demonstrates that the 'law' does not hold, and that is the end of the matter. The question for him is not whether there are rules in the former sense, nor whether we understand actions with reference to them. The question, rather, is whether there are *laws* covering the behaviour of societies, whether these laws are ultimate, and whether, in the social sciences, we are to be able to explain the workings of societies simply by reference to the behaviour of individual human beings. Even when tightened up, therefore, Mandelbaum's theses are not merely compatible with Methodological Individualism, but they are utterly irrelevant to questions concerning the latter's status. The fact that they could have been so much as considered to be relevant is due, in the end, to an equivocation on the word 'explain'.

The natural wish of the Methodological Individualist is to demonstrate the *logical impossibility* of Methodological Socialism—a wish, whose fulfilment would very nearly confer, upon his own position, the accolade of logical necessity. Watkins has attempted to demonstrate the logical impossibility of historical materialism, and while it is true that the refutation of historical materialism would no more demolish

Methodological Socialism than a refutation of Psychologism would demolish Methodological Individualism—since these are only *instances* of the main positions—the fact is that Watkins's argument could easily bear transfer to the more general position. The demonstration presupposes four distinct theses:

(α) There are predicates which range over social individuals.
(β) There are predicates which range over individuals.
(γ) It is a necessary condition for E to be an adequate explanation of a phenomenon e, that a sentence describing e be exhibited as a deductive consequence of premises.
(δ) There can be no non-logical term in the conclusion of a deductive argument which does not appear in that argument's premises.

Now, if there is to be an explanation E of some piece of individual behaviour e, the explanandum, i.e. the sentence which is used to formulate e, must employ predicates which range over individual human beings. Let S be such a sentence. The explanation will minimally require that S be exhibited as a deductive consequence (γ) of premises, and amongst these premises must appear at least one sentence which contains at least one predicate which ranges over individual human beings (δ). Accordingly, from premises containing sentences which employ *only* predicates ranging over social individuals, S cannot be deduced, and hence e cannot be explained. As Watkins puts it,

No description, however complete, of the productive apparatus of society, or of any other non-psychological factors, will enable you to deduce a single psychological conclusion from it, because psychological statements logically cannot be deduced from wholly non-psychological statements.[1]

And he adds:

That an explanation which begins by imputing some social phenomenon to human factors cannot go on to explain those factors in terms of some inhuman determinant of them, is a necessary truth.[2]

I do not in fact think it is a necessary truth, though I do agree that it is a necessary consequence of the theses (α)–(δ), but the chief difficulty with this, for the Methodological Individualist, is that *an exactly analogous argument will show that Methodological Individualism fails*. This argument is easily constructed by substituting, in the above proof, 'predicates which range over social individuals' for 'predicates which range over individual human beings', and vice versa. Thus the argument boomerangs. If it is cogent, and entails the impossibility of Methodological Socialism, it also entails the impossibility of Methodological Individualism. The wisest course at this point appears to be logical disarmament and peaceful philosophical co-existence.

On the other hand, the conflict between our methodologies, if we are now to assume that each of them is possible, can be pitched at another point. Let us accept the general correction of the four theses (α)–(δ), but consider in more detail precisely what is required if we are both to accept them and regard each of our methodologies as possible. It seems to me that the following adjustment must be made: an explanation, acceptable to Methodological Individualism, of the behaviour of some social individual, must employ, amongst its premisses, at least *one* sentence which employs as least one predicate ranging over social individuals. Similarly for Methodological Socialism. It must allow, amongst the premisses of *its* explanations, at least *one* sentence which employs at least *one* predicate which ranges over individual human beings. Not merely can this condition be met, but in meeting it, we will have something which seems to me a far more plausible way of representing either methodology than any we have so far discussed. The condition could be met if, amongst the premisses of the explanations, we had at least one law-like sentence of the following forms:

$$(L\text{-}1) \quad (x)\,(y)\,(P_i x \supset P_s y)$$
$$(L\text{-}2) \quad (y)\,(x)\,(P_s y \supset P_i x),$$

where P_i is a predicate which ranges over individual human beings, and where P_s is a predicate which ranges over social individuals, and where the antecedent is understood as describing a set of initial conditions for the state of affairs described by the consequent. Now any explanation, which included a law-like sentence of the form ($L\text{-}1$) amongst its premisses, would to that extent satisfy Methodological Individualism, for certain facts about the behaviour of individual human beings would then be initial conditions for certain facts pertaining for social individuals, and, for exactly similar consideration, an explanation containing, amongst its premisses, a law-like sentence of the form ($L\text{-}2$) would to that extent satisfy Methodological Socialism. Historical Materialism, incidentally, is precisely the claim that there are known laws of the form (*L*-2), in which certain facts about the prevailing productive systems are initial conditions for certain psychological facts pertaining for individual human beings.

Now we can give a more general interpretation of these law-like sentences. We can simply lay down the rule that the predicates in the antecedent range over a *different kind* of phenomena than the predicates in the consequent. Such law-like sentences then would be understood to describe causal connections between phenomena of different kinds; for example, that causal connections hold between certain brain-states of an individual, and certain psychological states of that same individual —assuming that 'brain-states' and 'psychological states' are instances of different kinds of phenomena. It is through the establishment of law-like sentences of this sort that we speak of *reduction* in a scientific sense. And it is important to stress that we can *only* speak of reduction, distinct from ordinary causal explanation, when we are dealing with essentially different kinds of phenomena. *Reduction* in this sense will mean: explaining one kind of phenomena with reference to phenomena of a different kind. This is *causal* reduction, of course, and must be distinguished from philosophical reduction, i.e. where a given set of terms is held to be translatable into another, and favoured set of terms. I have stressed that Methodological Individualism is *not* a thesis about philosophical reductionism, and is not concerned to demonstrate that the *meaning* of predicates which range over social individuals is to be rendered by means of predicates which range over individual human beings. But I am now saying that it is a thesis about *scientific* reduction; indeed it could not well subscribe to both kinds of reductionist programmes, for scientific reduction clearly presupposes that we are dealing with distinct kinds of phenomena: otherwise we could effect any reduction we chose by the simple device of making definitions. I am further suggesting that Methodological Socialism is also a thesis about scientific reduction, and if this construction of the matter is correct, Methodological Individualism is the claim that explanations employing laws of the form (*L*-1) are to be alone accepted as ultimate, or 'rock-bottom' explanations in the social sciences, and that laws of the form (*L*-2), and all so-called explanations which employ them, are to be rejected. Exactly the reverse of this will then be the claim of Methodological Socialism.

Having put the conflict in this form, we may once again ask whether any good reason may be found for choosing between Methodological Individualism and Methodological Socialism, and I shall say in advance that I am quite unable to find one. Never the less, a few words more might profitably be devoted to analysing the conflict in the form it has taken, and so to suggesting why it is hard to find good reasons for deciding it one way or another.

The concept of reduction in science has been analysed in considerable detail by Ernest Nagel, and in what follows I shall assume the essential correctness of his analysis,[1] and shall refer the reader to his work for arguments I shall not seek to summarize here. He is, of course, not responsible for the following remarks.

Suppose we were to speak of macro- and of micro-properties of societies. As a model, one might think of individual human beings as standing to societies in something like the relation in which molecules stand to gases. This might ultimately prove a misleading analogy, but for the moment it will be convenient to employ it to raise some problems. Now, in the case of gases, we speak of reduction in the following sense: some relationships and proportions between some macro-properties of gases are explained with reference to some relationships and proportions between molecules, for example variations in the temperature of a gas are explained with reference to variations in the mean kinetic energy of molecules in random motion, and hence with reference to the mechanical behaviour of aggregates of molecules. I suppose the Methodological Individualist, so far as he would accept this analogy, would speak of reduction in an exactly similar sense: relations between properties of societies, considered macroscopically, are to be explained with reference to relations between individual human beings, i.e. to societies considered 'microscopically'.

Now before the reduction of thermodynamics to mechanics was achieved, in the nineteenth century, scientists had reasonably clear criteria for applying the terms of the former theory to gases; reasonably clear laboratory procedures for measuring variations in the properties denoted by these terms, and hence reasonably clear ways of confirming, or disconfirming, law-like sentences intended correctly to describe relationships of co-variation amongst these properties. All of this was retained after the reduction was effected. All the known laws of thermodynamics were preserved. What had been achieved (in unspeakably rough terms) was this: the behaviour of gases, as described by thermodynamics, had been explained with reference to the behaviour of molecules, as described by mechanics, and so one might, if one wished, speak of mechanics as the more 'ultimate' of the two theories, and of molecules as 'the ultimate constituents' of gases. In general, we could speak as follows: if T-1 is a theory which explains and predicts the behaviour of phenomena P-1, and T-2 is a theory which explains and predicts the behaviour of phenomena P-2, then, if T-2 is reduced to T-1, in the sense that we can now also explain and predict the behaviour of P-2 with reference to P-1, T-1 is a more ultimate theory than T-2. Or, to revert to Watkins's language, T-2 may be considered a half-way theory for P-2, and T-1 a 'rock-bottom' theory for P-2. This does not, of course, mean that either P-2's are not 'real' or that sentences about P-2's are 'really about' P-1's.

Now let us suppose there were a theory of society comparable to thermodynamics, i.e. a theory which is concerned with societies in a macroscopic way. Imagine that such terms as 'social density' and 'cultural-economic elasticity' were among the terms of this theory, and that there were well-confirmed laws like 'Social density varies inversely as the cube of cultural economic elasticity'. *Should there be* such a theory, the Methodological Individualist would say: one can ultimately explain the variations in these properties, one can ultimately account for co-variations between them, by means of a theory concerned with the behaviour of individual human beings. To this theory we shall be able ultimately to reduce the macro-theory of society. But, of course, for us to speak of reductions in the scientific sense, it is first required that we have two theories. Of course, one *might* interpret Methodological Individualism as a negative injunction *against* seeking macro-theories of society. If so, it becomes extremely difficult to see what could now be meant by 'reduction' or by 'individuals being ultimate'. Moreover, by the Methodological Individualist's own admission, there are properties of societies which are different altogether from the properties of individuals. Are these properties to be left unaccounted for ? Surely the whole point of Methodological Individualism is that all relations and proportions between such properties are to be explained, ultimately, by reference to the behaviour of individual human beings. So the Methodological Individualist would find it self-defeating to adhere to such a negative injunction. When Watkins says, of Methodological Individualism, that it is a regulative principle, part of whose purpose is to discourage research in certain directions, he places himself in the ironic position of making the viability of his methodological programme dependent upon the success of exactly the sort of research it is calculated to discourage, and we can complete this uncanny dialectic by making exactly similar points about the Methodological Socialist.

But let us suppose that the social sciences advance, and that one day a reduction is achieved which is exactly of the sort to bring joy to the Methodological Socialist's heart! The behaviour of *individual human beings* really is explained with reference to large-scale variations in the macro-properties of societies! The result would be in many ways the reverse of the example of gases. Not merely would the direction of explanation—*von oben bis unten*—be different, but the operational problems would be different as well. For while the macro-properties of gases lie within the realm of observables, the micro-properties do not. Conversely, the behaviour of human individuals lies within the realm of observables, while it has long since been agreed that we do not observe the workings of societies as such. So it is with this difference that I shall now concern myself.

It is a commonplace consideration, I should think, that when a causal theory is advanced in the sciences, asserting, say, that changes in a dependent variable v-1 are explained and predictable by reference to changes in an independent variable v-2, it is necessary that we have some independent means of measuring variations in these two variables. The Boyle–Charles law is thus confirmable by observations on pressure-gauges and thermometers. Yet suppose the *sole* way in which we could assign a value to the mean kinetic energy of unobservable molecules was by measuring the temperature of an observable body of gas. One might begin to wonder about the status of a law which held that variations in temperature were explained and predictable by reference to variation in the mean kinetic energy of gas molecules. Imagine someone were to assert that the motion of an eel through water is due to the simultaneous motion of millions of sub-microscopic limbs connected to the eel's side, like an immense galley. As the rotational frequency of the unobservable oars increases, the eel moves faster in proportion, but unhappily, the only way in which we can assign a value to the 'mean rotational frequency' of the 'ultimate constituents' of the eel would be by measuring the rate at which the eel moves through the water. Scientists might decently reject the entire account as a fairy-tale, but in fact the reduction of thermodynamics to mechanics did not consist in asserting a simple connection of co-variation between the micro- and macro-properties of gases. The relation was far more complex, and mechanics itself had a long and distinguished history before it was modified and extended to account for thermal phenomena. The fact is that there are various ways in which we can assign values to the mean kinetic energy of molecules in random motion, but, of course, never by directly applying yardsticks to molecules. These remain beyond observation, and our relations to them are indirect, and via elaborate pieces of laboratory and mathematical apparatus. Nevertheless, it remains possible to raise doubts about molecules. A radical empiricist retains the option of trying to eliminate reference to them in favour of a reconstructed theory which makes use only of observational predicates. Or one may regard them, as Professor Marrou regards those '*abstraits socio-culturels*', as concepts only, though of an unquestioned use for the organization of experience. But these options have to do with issues in ontology and in meaning, and it will be common ground that the theory, meanwhile, has an uncontested explanatory and predictive power.

Now precisely such options would remain with the sort of reductive theory I am supposing. The radical empiricist and the instrumentalist alike can specify interpretations of the laws and language of this theory of an exactly analogous sort to those which they would specify in the kinetic theory of heat. In point of unobservability, there would be little for them to choose between molecules and organic social systems. Their positions, in either case, are quite independent indeed from that of the Methodological Individualist, whose claim, only, is that such a theory is not 'ultimate', even if, *pro tempore* it might be supposed to have genuine predictive and explanatory power. His objection certainly cannot be based on ontological grounds, nor upon the sort of reservations concerning meaningfulness which animates the radical empiricist. But what is it exactly that his objections *are* based on? What exactly would be wrong with such a theory? It is this I find very hard to determine. I *think* I know what he *fears* such a theory would entail. I shall turn to this in a moment. But I should like, before doing so, to point out how very little philosophical interest remains in this controversy. The philosophical interest it might *appear* to have derives from the other, interesting, issues which resemble it. Once isolated from these, it quickly degenerates into a quite unphilosophical preoccupation with science-fiction. But it is a *dangerous* preoccupation, for it might become a self-fulfilling thesis. Social scientists, persuaded in advance that a certain sort of theory is suspect, might abandon the quest for such theories. And this might, in the end, have consequences of exactly the sort feared by the Methodological Individualist.

What the Individualist fears, I submit, is this. He feels that theories of the sort he wishes to reject would entail that we do not hold our destinies in our own hands, that we are, as it were, dragged along by the development of the social individuals we are part of, and which have a life of their own.[1] Now even if this were entailed by such theories, it would be wholly contrary to the spirit of Philosophy to refuse to look for such theories. We may sympathize with, but scarcely approve of the attitude which, according to legend, led Pythagorians to do away with the man who discovered that there is no rational number, the square of which is equal to 2. On the other hand, it seems to me that the theories in issue entail no such thing, and an important step towards seizing our destinies in our own hands might be taken with the discovery of exactly these sorts of theories. It does not follow, from the fact that we successfully explain the gross, thermal behaviour of gases with reference to the mechanical behaviour of molecules, that we cannot control the mechanical behaviour of molecules. By applying gross heating and cooling apparatuses to bodies of gas, we may modify the value of the mean kinetic energy of the molecules contained in it, but, by symmetrical analogy, should we ever be able to explain the behaviour of individual human beings with reference to the behaviour of large-scale processes in *social* individuals, nothing would prevent us from similarly *controlling* those large-scale processes by operating at the 'micro-level', i.e. upon individual human beings. Science is not noted for *diminishing* our control over things.

Never the less, even were we to assume the correctness of methodological individualism in assuming that social processes are but the complicated outcome of individual actions, the extent to which we do have control over our own destinies is in some measure limited. For usually, I have argued, the outcomes of our actions is seldom intended by us, and actors, unless they survive, and know retrospectively, are blind to the significance of their actions because they are blind to the future. We see this nowhere more clearly than in the sort of historical account at the beginning of this chapter. The changes Miss Wedgewood described were 'insensible' and few were aware they were taking place. This, not because a more delicate instrument could have detected them, for no such instrument could be built: the changes could not have been detected at the time, for it is only in the light of future events that they could so much as have been described. To have been able to control or modify these changes would accordingly have required that men see their own actions in a perspective unfortunately not available to them, the perspective, namely, of historians future to their actions. Not knowing how our actions will be seen from the vantage point of history, we to that degree lack control over the present. If there is such a thing as inevitability in history, it is not so much due to social processes moving forward under their own steam and in accordance with their own natures, as it is to the fact that by the time it is clear what we have done, it is too late to do anything about it. 'The owl of Minerva takes flight only with the falling of the dusk.' Philosophies of history attempt to capture the future without realizing that if we knew the future, we could control the present, and so falsify statements about the future, and so such discoveries would be useless. We capture the future only when it is too late to do anything about the relevant present, for *it* is then past and beyond our control. We can but find out what its significance was, and this is the work of historians: history is made by them.

NOTES

1 Karl Marx and Friedrich Engels, *The German Ideology* (New York; International Publishers, 1947), p. 22. 'As soon as labor is distributed, each man has a particular, exclusive sphere of activity, which is forced upon him and from which he cannot escape. He is a hunter, a fisherman, a shepherd, or a critical critic, and must remain so if he does not want to lose his means of livelihood; while in communist society, where nobody has one exclusive sphere of activity but each can become accomplished in any branch he wishes, society regulates the general production and thus makes it possible for me to do one thing today and another tomorrow, to hunt in the morning, fish in the afternoon, rear cattle in the evening, criticize after dinner, just as I have a mind, without ever becoming hunter, fisherman, shepherd, or critic.' Marx's reluctance to speak in any detail about the classless society was consonant, of course, with his general theory that forms of life and consciousness reflect the material conditions of existence ('the production of ideas, of conceptions, of consciousness, is at first directly interwoven with the material activity and the material intercourse of men, the language of real life': *ibid.* pp. 13–14)—so how is one to speak of the 'ideas, conceptions, etc.' which *will* exist under a form of material existence which has never yet existed? Moreover, in the classless society, men are at any rate to be liberated from these material causes, and free to control their lives. So one can only say that at that time things will be 'the opposite' of what they are now, so at best a kind of negative characterization is possible. But it is not easy to positively identify what 'not-*A*' is to designate. Cf. Engels, 'The Origins of Family, Private Property, and the State', in Marx and Engels, *Selected Works* (London; Lawrence & Wishart, 1950), II, 219: 'What we can conjecture at present about the regulation of sex relations after the impending effacement of capitalist production is, in the main, of a negative character, limited mostly to what will vanish.'

2 'If all contradictions are once for all disposed of, we shall have arrived at so-called absolute truth—world history will be at an end. And yet it has to continue, although there is nothing left for it to do—hence, a new, insoluble contradiction.' (Friedrich Engels, 'Ludwig Feuerbach and the End of Classical German Philosophy', in Marx and Engels, *Selected Works*, II, 330.) To be sure, Engels here is speaking of Hegel, but in fact the same 'contradiction' holds in his own system. In the classless society, or for post-revolutionary history, the Marxist theories of history will not have application. See the following note.

3 History, apparently, in the Marxist view, admits of a theory only so far as human beings are driven by forces over which they have no control. But in the classless society men will be free from historical forces, and hence 'make their own' history, instead of being 'made by it'. Thus 'The whole sphere of conditions of life which environ man, and which have hitherto ruled man, now comes under the dominion and control of man, who for the first time becomes the real, conscious lord of Nature, because he has now become master of his own social organization. ... Man's own social organization, hitherto confronting him as a necessity imposed by Nature and history, now becomes the result of his own free action. The extraneous objective forces that have hitherto governed history pass under the control of man himself. Only from that time will man himself, more and more consciously, make his own history. ... It is the ascent of man from the kingdom of necessity to the kingdom of freedom.' Friedrich Engels, 'Socialism: Utopian and Scientific' in Marx and Engels, *Selected Works*, II, 140–1.

PAGE 4

1 Immanuel Kant, 'Idea of a Universal History from a Cosmopolitical Point of View', tr. by W. Hastie, in Patrick Gardiner (ed.), *Theories of History* (Glencoe; Free Press, 1959), p. 23.

PAGE 7

1 Karl Löwith, *Meaning in History* (Chicago; University of Chicago Press, 1957), p. 1. Cf. 'What projects customarily referred to as "philosophies of history" have in common is the aim of giving a comprehensive account of the historical process that "makes sense".' Patrick Gardiner, Introduction in *op. cit.* p. 7.

PAGE 8

1 'Here I should like to say: a wheel that can be turned though nothing else moves with it, is not part of the mechanism [Maschine].' Ludwig Wittgenstein, *Philosophical Investigations*, tr. G. E. M. Anscombe (New York; Macmillan, 1953), para. 271.

PAGE 9

1 Löwith, *op. cit.* p. 1. I cannot accept Löwith's *reasons* for saying this, however, which seem rhetorical.
2 The distinction between prediction and prophecy I borrow from Karl Popper. See his 'Prediction and Prophecy in the Social Sciences', in Gardiner, *op. cit.* pp 276 ff. By 'prophecy', Popper means an unconditional

prediction. He allows only conditional predictions (i.e. given condition *C*, then *E*), or predictions derived from these. He argues that historicists not only give unconditional predictions, but give them for systems where it is not legitimate to do so. Unconditional predictions are licit only when derived from conditional ones, and then with respect to 'well-isolated, stationary, and recurrent systems'. Society is 'open', however. This is not quite the sense I am giving to the notion of prophecy, as will appear. Nor do I find historicism quite so illegitimate as Popper does, here and elsewhere in his writings. Cf. especially *The Poverty of Historicism* (Boston; Beacon Press, 1957), ch. II and *passim*. I deal in part with this in ch. XII.

3 E.g. Hitler, who was given to such utterances as 'The war is won', spoken in the early 40's. Hitler's confident descriptions of the present in the light of a future he gave every appearance of having a special insight into must have accounted, in some measure, for the remarkable power he had over people's minds.

4 Donald Williams, 'More on the Ordinariness of History', *Journal of Philosophy*, LII, no. 10, p. 272.

PAGE 13

1 W. H. Walsh, '"Meaning" in History', in Gardiner, *op. cit.* pp. 296 ff.

PAGE 14

1 G. W. F. Hegel, *The Philosophy of History*, tr. J. Sibree (New York; Willey Book Co., 1944), p. 350. 'We find, moreover, in the east of Europe, the great *Sclavonic* nation. . . . These people did, indeed, found kingdoms and sustain spirited conflicts with the various nations that came across their path. Sometimes as an advanced guard—an intermediate nationality—they took part in the struggle between Christian Europe and unchristian Asia. The Poles even liberated beleaguered Vienna from the Turks; and the Sclaves have to some extent been drawn within the sphere of Occidental Reason. Yet this entire body of peoples remains excluded from our consideration, because hitherto it has not appeared as an independent element in the series of phases that Reason has assumed in the World. Whether it will do so hereafter, is a question that does not concern us here; for in History we have to do with the past.' Again: 'We have confined ourselves to the consideration of the progress of the Idea, and have been obliged to forego the pleasure of giving a detailed picture of the prosperity, the periods of glory that have distinguished the career of people, the beauty and grandeur of the character of individuals, and the interest attaching to their fate in weal and woe. Philosophy concerns itself only with the glory of the Idea mirroring itself in the History of the World.' (*Ibid.* p. 457.)

PAGE 15

1 Jacob Burkhardt, *Force and Freedom: Reflections on History*, tr. J. H. Nichols (New York; Pantheon Books, 1943), p. 80.
2 *Ibid.*

PAGE 19

1 Thucydides, *The Peloponnesian War*, tr. by Crawley (New York; The Modern Library, 1934), bk. 1, 1. 'Indeed this was the greatest movement yet known in history, not only of the Hellenes, but of a large part of the barbarian world—I had almost said of mankind. For though the events of remote antiquity, and even those that more immediately preceded the war, could not from lapse of time be clearly ascertained, yet the evidences, which an inquiry carried back as far as was practicable leads me to trust, all point to the conclusion that there was nothing on a great scale, either in war or in other matters.' This statement has been radically misinterpreted by Spengler. Thucydides' 'lack of historical feeling', Spengler writes, 'is conclusively demonstrated on the very first page of his book by the astounding statement that before his time (about 400 B.C.) no events of importance had occured in the world!' (Oswald Spengler, *The Decline of the West*, tr. by C. F. Atkinson: New York; Knopf, 1946, 1, 10.) Thucydides says nothing of the sort, but only that on the basis of the best evidence he knew of, nothing as large-scale as the war between Sparta and Athens had occurred. Spengler says that 'what is absolutely hidden from Thucydides is perspective, the power of surveying the history of centuries, that which for us is implicit in the very conception of an historian.' (*Ibid.*) To be sure, what Thucydides lacked was a predecessor as gifted as himself. And his painstaking accuracy, which even Spengler admires, was meant, as we shall see, for 'all time' so that men might always thenceforth be able to *use* his work. 'Someone said: "The dead writers are remote from us because we *know* so much more than they did." Precisely, and they are that which we know.' T. S. Eliot, 'Tradition and Individual Talent', in *Selected Essays: 1917–1932* (New York; Harcourt Brace, 1932), p. 6.

2 Thucydides, bk. 1, xxii. 'With reference to the narrative of events, far from permitting myself to derive it from the first source that came to hand, I did not even trust my own impressions, but it rests partly on what I saw myself, partly on what others saw for me, the accuracy of the report being always tried by the most severe and detailed tests possible. My conclusions have cost me some labor from want of coincidence between accounts of the same occurrences by different eye-witnesses, arising sometimes from imperfect memory, sometimes from undue partiality for one side or the other.' The celebrated description of the plague is a fair example of this: 'I shall simply set down its nature, and the symptoms by which it perhaps may be

recognized by the student, if it should ever break out again. This I can the better do, as I had the disease myself, and watched its operation in the case of others.' (Bk. II, xlviii.) This, incidentally, illustrates as well the sort of *use* to which Thucydides felt his work might in general be put. See the following note.

PAGE 20

1   'The absence of romance in my history will, I fear, detract somewhat from its interest; but if it be judged useful by those inquirers who desire an exact knowledge of the past as an aid to the interpretation of the future, which in the course of human things must resemble if it does not reflect it, I shall be content. In fine, I have written my work, not as an essay which is to win the applause of the moment, but as a possession for all time.' (Bk. II, xxii.)

2   τοιαῦτα καὶ παραπλήσια: 'such and such-*like*'. Liddell and Scott, *Greek-English Lexicon* (New York; Harpers, 1848), p. 1110. That is to say, 'resemble or *exactly* resemble'.

3   'His method was inductive. He cited facts and then drew conclusions from them. He believed in cycles of history and he wished to aid the cause of civilization by showing men how, under a given set of circumstances, individuals, and above all communities, had in the past acted rightly or wrongly, in order that in the future the mistakes of the past might be avoided.' G. B. Grundy, *Thucydides and the History of his Age* (London; John Murray, 1911), p. 8.

PAGE 21

1   See note 1, p. 19. Also: 'The Median war, the greatest achievement of past times, yet found a speedy decision in two actions by sea and two by land. . . . Never had so many cities been taken and laid desolate . . . never was there so much banishing and bloodshedding, now in the field of battle, now in the strife of action. Old stories of occurrences handed down by tradition, but scantily confirmed by experience, suddenly ceased to be incredible; there were earthquakes of unparalleled extent and violence. . . .' (Bk. I, xxiii.)

PAGE 22

1   For instance, his initial emphasis upon the great scale of the war may have been by way of advertising the subject of his history as 'greater and more interesting than that of his predecessor, Herodotus'. (Grundy, *op. cit.* p. 3.) Cf. also the individious contrast with the Persian War, cited in n. 1, p. 21.

PAGE 23

1   Richard Taylor, 'Fatalism', *Philosophical Review*, LXXI, 1 (January, 1962), 56–66.

2 And of course if it *did* have its intended use, the future would *not* have resembled the past. By marking the pockets of quicksand where people have been trapped, one expects the future to be empty of victims of at least these pockets.

3 'We can at least conceive of a change in the course of nature; which sufficiently proves that such a change is not absolutely impossible. To form a clear idea of anything is an undeniable argument for its possibility, and is alone a refutation of any pretended demonstration against it.' (David Hume, *A Treatise of Human Nature*, bk. I, part III, sect. vi.) Imaginability is Hume's criterion of logical possibility. If the 'opposite' of a state of affairs *S* is imaginable, *S* is not necessary. So we cannot *demonstrate* that there will be no such change. Nor can we appeal to experience, since this quite begs the question. But we have only demonstration and experience as bases for judgement. Hume's criteria are far too confused to untangle here.

PAGE 30

1 The first philosopher to have held this view, or one closely connected to it, seems to have been Peirce. 'It cannot be denied', he writes, 'that acritical inferences may refer to the Past in its capacity as Past; but according to Pragmaticism, the conclusion of a reasoning power must refer to the Future. For its meaning refers to conduct, and since it is a reasoned conclusion must refer to deliberate conduct, which is controllable conduct. But the only controllable conduct is future conduct. . . . Thus, a belief that Columbus discovered America really refers to the Future.' (C. S. Peirce, *Collected Papers*, edited by C. Hartshorne and P. Weiss: Cambridge, Mass.; Harvard University Press, 1934, V, para. 461.) Again: 'The truth of the proposition that Caesar crossed the Rubicon consists in the fact that the further we push our archeological and other studies, the more strongly will that conclusion force itself on our minds forever—or would so if study were to go on forever.' (*Collected Papers*, V, para. 544.)

2 Russell's view was that every meaningful proposition must be either true or false. His celebrated Theory of Descriptions was specifically engineered to handle sentences whose meaning was plainly understood, but which could not readily be assigned a truth-value, for (1) these sentences seemed to require that there actually exist something for their subject-term to refer to but (2) no such thing existed. Rather than manufacture special entities for such sentences as 'The present King of France is bald' to be *about*, he recast them in such a manner as to require no new entities and to enable one to assign the value 'false' to that sentence and its natural contradictory, meanwhile preserving the Principle of Contradiction. In general, all sentences which employ a singular referring expression as a subject term, and which in fact have no referendum, are *false*. See especially B. Russell,

Introduction to Mathematical Philosophy, 2nd edition (London; Allen and Unwin, 1920), ch. XVI. Few pieces of recent philosophical analysis have been more heatedly discussed, and indeed the entire recent history of Anglo-American philosophy could be written with specific reference to the Theory of Descriptions. The chief critical attack is due to P. F. Strawson, in his 'On Referring', *Mind*, LIX (July, 1950). The subsequent literature is considerable.

3 I distinguish between '*a* doubts that *p*' and '*a* is sceptical with regard to *p*'. The first implies that *a* believes that not-*p*, while the second implies that *a* has no grounds for choosing between *p* and not-*p*, and suspends belief as between them.

PAGE 32

1 Margaret MacDonald, 'Some Distinctive Features of Arguments used in Criticism of the Arts'. Reprinted in M. Weitz (ed.), *Problems in Aesthetics* (New York; Macmillan, 1960), p. 696.

PAGE 33

1 F. Nietzsche, *Beyond Good and Evil*, II, para. 68. The citation by Freud is in 'The Psychopathology of Everyday Life', in *Basic Writings* (New York; Modern Library, 1938), p. 103.

PAGE 34

1 See C. I. Lewis, *Mind and the World Order* (New York; Dover, 1956), pp. 148–53. Lewis takes the problem up again in *An Analysis of Knowledge and Valuation* (Lasalle, Ill.; Open Court, 1946), pp. 197–200. For a discussion of some of the problems, see Evelyn Masi, 'A note on Lewis's Analysis of the Meaning of Historical Statements', *Journal of Philosophy*, XLVI, 21 (1949), 670–4; and Israel Scheffler, 'Verifiability in History: A Reply to Miss Masi', *Journal of Philosophy*, XLVII, 6 (1950), pp. 158–66. Scheffler makes the important point that Lewis's analysis is concerned with 'intensional meaning' rather than with 'objective reference', but it is not clear to me that Lewis *himself* is aware of the distinction, or that, if he were aware of it, he could readily accommodate it to his analysis. And if we drop 'objective reference', what of 'future'?

PAGE 35

1 *Mind and the World Order*, p. 140.
2 *Ibid.* p. 142.
3 Or, as he later put it, 'translated into'. Cf. *An Analysis of Knowledge and Valuation*, ch. VII, *passim*, and especially pp. 182–5.

PAGE 36

1 *Mind and the World Order*, p. 149. Cf. *An Analysis of Knowledge and Valuation*, p. 197: 'By depicting the meaning of "Caesar died" as consisting in what would verify it to us in future possible experience, this conception may be charged with translating what is past into something which is exclusively future.'

PAGE 37

1 John Dewey, 'Realism without Monism or Dualism: I. Knowledge Involving the Past', *Journal of Philosophy*, XIX, 12 (1922), 314. This is Dewey's fullest statement of what we can recognize as a general Pragmatist thesis. Critics of Pragmatism such as Blanshard, Lovejoy, and Santayana never fail to stress it as a vulnerable point. For a recent discussion, see Richard Gale, 'Dewey and the Problem of the Alleged Futurity of Yesterday', *Philosophy and Phenomenological Research*, XXII, 4, pp. 501 ff.

2 John Dewey, *The Influence of Darwin on Philosophy* (New York; Holt, 1910), pp. 160–1.

PAGE 38

1 *Mind and the World Order*, p. 151.

PAGE 39

1 *Mind and the World Order*, p. 151.
2 *Ibid.* p. 153.

PAGE 40

1 David Hume, *A Treatise of Human Nature*, bk. I, part I, sect. iii: 'The ideas of memory are much more lively and strong than those of the imagination, and the former faculty paints its objects in more distinct colours than any which are employed by the latter.' Cf. *op. cit.* part III, sect, v. Hume later employed exactly the same criterion for distinguishing between 'ideas' and 'impressions', in order to answer the question: How can I tell whether I am perceiving an x or only just thinking about an x? (See his *Inquiry Concerning Human Understanding*, sect. ii.)

2 R. F. Holland, 'The Empiricist Theory of Memory', *Mind*, LXIII, 252 (1954), 466.

3 Bertrand Russell, *The Analysis of Mind* (London; Allen and Unwin, 1921), p. 162. This is not the sole criterion Russell offers: he also mentions 'amounts of context'. Notice that 'feelings of pastness' are supposed to differentiate between near and remote memories: 'feelings of familiarity' are to differentiate between memories *überhaupt* and imaginings.

Whatever the case, Russell's attempts are dominated by accepting as valid the essential question which Hume was concerned with, viz.: 'It seems that there must be some mark or sign whereby a remembering state of mind can be distinguished from an imagining one. So that one proceeds to ask: What is this mark or sign?' (Holland, *loc. cit.* p. 465.) And this is Lewis's problem as well.

4 Bruce Waters, 'The Past and the Historical Past', *Journal of Philosophy*, LII (1955), 253–64.

5 See for example *An Inquiry into the Forgery of the Etruscan Terracotta Warriors in the Metropolitan Museum of Art*, The Metropolitan Museum of Art Papers, no. 11 (1961).

6 *Mind and the World Order*, p. 150. Cf. *An Analysis of Knowledge and Valuation*, p. 200, where a similar recourse to challenge is made.

PAGE 41

1 For example, Bertrand Russell, *partout*.

PAGE 45

1 A. J. Ayer, *Language, Truth and Logic*, 2nd edition (London; Gollanz, 1946), p. 102.

2 *Ibid.*

3 A. J. Ayer, *The Problem of Knowledge* (Penguin Books, 1956), p. 154.

PAGE 46

1 A. J. Ayer, Preface to the Second Edition, *Language, Truth, and Logic*, p. 119.

PAGE 47

1 A. J. Ayer, *The Foundations of Empirical Knowledge* (London; Macmillan, 1940), pp. 167–8.

PAGE 48

1 Ludwig Wittgenstein, *Tractatus Logico-Philosophicus* (London; Kegan Paul, 1922), 6.4311.

PAGE 49

1 Preface to Second Edition, *Language, Truth, and Logic*.

PAGE 51

1 For difficulties in this notion, see John Hospers, *An Introduction to Philosophical Analysis* (New York; Prentice Hall, 1953), p. 442.

PAGE 53

1 A. J. Ayer, *The Problem of Knowledge*, p. 160.

2 *Ibid.*

PAGE 54

1 But we might, for example, have 'spases' as well as tenses in our language. Thus, in learning the conjugation of verbs, students would be required to master 'the right-hand present indicative', 'the left-hand future', and so on. We might then have a difficulty in giving a spatially neutral translation of a language so inflected. In fact, we have devices for giving any information spases might allow, just as languages without tenses (for example, Chinese) are able to give information concerning the temporal direction of a reference. English has, in strict grammatical fact, but two distinct tenses. But I shall later show how a good deal of our vocabulary is temporal, quite apart from tenses.

PAGE 55

1 Ayer, *op. cit.* p. 160.

PAGE 58

1 See A. N. Prior, *Time and Modality* (Oxford; Clarendon Press, 1957), p. 9.
2 See ch. VIII.

PAGE 59

1 Ayer, *op. cit.* p. 161.

PAGE 66

1 The fastidious may baulk at the suggestion that there are false memories. If what I claim to remember did not happen, then I simply do not remember it. Hence I do not have a memory of it. According to this view, it is analytic that if *a* remembers *E*, then *E* in fact happened. So it may be. This, however, but transfers the problem; the question being now which of what seem to be memories are so in fact. It is this question the sceptic wishes to say that we cannot answer, and cannot, in particular, answer on the basis of a simple examination of our seeming-memories. But then what alternative have I? I plainly cannot examine what these purport to be memories *of*. At all events, I shall hope that my use of the expression 'false memory' is not misleading.
2 H. H. Price, *Thinking and Experience* (Cambridge, Mass.; Harvard University Press, 1953), p. 84.

PAGE 77

1 David Hume, *A Treatise of Human Nature*, bk. I, part III, sect. 3.
2 Bertrand Russell, *The Analysis of Mind* (London; Allen & Unwin, 1921), pp. 159–60.

PAGE 78

1 That is, when 'knows' takes the name of an individual as its accusative. It is then what I call *existence-entailing*. Comparably, '*a* knows that *p*' is *truth-entailing*, for from it the truth of *p* follows logically.

2 But notice that 'is called such-and-such' is a temporally neutral predicate, e.g. in roughly the same way that 'seems to be such and such' is what one might call *existence-neutral*. It does not commit one in the way in which '*is* such and such' does. '*X* is called a father' is compatible with '*X* is not a father'.

3 Bertrand Russell, *op. cit.* p. 160.

4 'But on the contrary, the notion that the world came into being five minutes ago, complete with a population which 'remembered' a wholly unreal past, is fascinating—but untenable.' R. J. Butler, 'Other Dates', *Mind*, LXVII, no. 269 (1959), p. 16.

PAGE 79

1 The literature on this is considerable. For the best recent discussions, see especially C. G. Hempel, 'The Theoretician's Dilemma' in Feigl, Scriven, and Maxwell (eds.), *Concepts, Theories, and the Mind-Body Problem*, Minnesota Studies in the Philosophy of Science, vol. II (University of Minnesota Press, 1958); and Israel Scheffler, 'Prospects of a Modest Empiricism', *Review of Metaphysics*, x, nos. 3 and 4 (1958).

PAGE 80

1 With the exception of Peirce, whose views on the philosophy of history have yet to be carefully studied. Thus: 'When I say a reductive inference is not a matter for belief at all, I encounter the difficulty that there are certain inferences which scientifically considered are undoubtedly hypotheses, and yet which practically are perfectly certain. Such, for instance, is the inference that Napoleon Bonaparte really lived at about the beginning of this century, a hypothesis which we adopt for purposes of explaining the concordant testimony of a hundred memoirs, the public records of history, traditions, and numberless monuments and relics. It would surely be downright insanity to entertain a doubt about Napoleon's existence.' Nevertheless, the latter is 'quite aside from the purpose of science.... It is extra-scientific'. C. S. Peirce, *Collected Papers*, v, para. 589.

2 Cf. 'Theoretical statements offer an explanation of the facts, but not in terms of more facts.' P. Herbst, 'The Nature of Facts', in A. Flew (ed.), *Essays in Conceptual Analysis* (London; Macmillan, 1956).

PAGE 81

1 It is of course not quite fair to Instrumentalism to regard it as a kind of refuge

to which we may retreat when driven back to sceptical attack. It was originally offered, and is surely currently defensible, as a positive theory in its own right, and not as a kind of fall-out shelter.

2 The suggestion is taken from Richard Taylor, 'The "Justification" of Memories and the Analogy of Vision', *Philosophical Review*, LXV, no. 2 (1956), p. 198.

PAGE 82

1 Butler, 'Other Dates', p. 16.
2 Which is precisely the move that Lewis made. All statements purportedly asserting something of physical objects are to be translated out into sets of conditionals involving actions and experiences.

PAGE 83

1 And this statement would hardly be accepted on even an Instrumentalist criterion, for it serves rather poorly to organize the present very coherently.

PAGE 84

1 Gilbert Ryle, *The Concept of Mind* (New York; Barnes & Noble, 1949), pp. 149–53 and *passim*.

PAGE 88

1 Charles Beard, 'Written History as an Act of Faith', *The American Historical Review*, XXXIX, 2 (1934), p. 219. Reprinted in Hans Meyerhoff (ed.), *The Philosophy of History in Our Time* (New York; Anchor Books, 1959), p. 140. Page references are to the latter.
2 *Ibid.*
3 Beard seems not to have felt it to be an especially ordinary way of looking at history: 'Although this definition of history may appear, at first glance, distressing to those who have been writing lightly about "the science of history", and "the scientific method" in historical research and construction, it is in fact in accord with the most profound contemporary thought about history ...' (*ibid.*). This is remarkable both for Beard's notion that his characterization is profound and that history, so characterized, is somehow incompatible with any use of 'scientific method'.

PAGE 89

1 *Ibid.*
2 R. Butler, 'Other Dates', p. 32.

PAGE 92

1 See the brilliant paper of David Pears, 'Time, Truth, and Inference', in *Proceedings of the Aristotelian Society*, vol. LVI. Reprinted as ch. XI in A. Flew (ed.), *Essays in Conceptual Analysis*.

2   Cf. A. J. Ayer, *The Problem of Knowledge*, p. 152. Much the same considerations apply, I think, to his *deictic* definition of 'is present'. 'The notion of the present . . . may be defined ostensively . . . as the class of events contemporary with *this*, where *this* is any event that one chooses to indicate at the given moment.'

1   Charles Beard, 'That Noble Dream', *The American Historical Review*, XLI, 1 (1935), pp. 74–87. Reprinted in Fritz Stern (ed.), *The Varieties of History* (New York; Meridian Books, 1956), p. 323. Page references are to the latter. Note that the word 'objectively' in this context contrasts with 'sees through a medium' and so means, roughly, 'perceives directly'. More particularly, Beard means that one does not have, in science, to *infer* propositions about one's subject-matter on the basis of what one does perceive, for one perceives the subject-matter as such. And with history it is quite otherwise.

1   I refer here, of course, to the important discussion of 'seeing as' and 'aspect blindness' in L. Wittgenstein, *Philosophical Investigations*, II, xi. See also N. R. Hanson, *Patterns of Discovery* (Cambridge University Press, 1958), especially ch. 1.
2   Beard, *loc. cit.* p. 324.

1   Claude Bernard, *Introduction à la Médecine Expérimentale* (Paris; 1865), p. 67. Cited by Pierre Duhem, in Philip Wiener (tr.), *The Aim and Structure of Physical Theory* (Princeton; Princeton University Press, 1954), pp. 181–2.
2   Duhem, *op. cit.* p. 182.
3   Beard, *loc. cit.* p. 324.

1   'Thermodynamics can be reduced to a mechanics that post-dates 1866, but it is not reducible to a mechanics as this science was conceived in 1700. Similarly, a certain part of chemistry is reducible to a post-1925 physical theory, though not to a physical theory of a hundred years ago.' Ernest Nagel, 'The Meaning of Reduction in the Natural Sciences', in R. Stauffer (ed.), *Science and Civilization* (Madison; University of Wisconsin Press, 1949). Reprinted in A. Danto and S. Morgenbesser (eds.), *Philosophy of Science* (New York; Meridian Books, 1960), p. 307.

1   Beard, *loc. cit.* p. 324

PAGE 101

1 Duhem, *op. cit.* p. 183.

2 The best recent discussion is in C. F. Presley, 'Francis Bacon: His Method and His Influence', *The Australian Journal of Science*, XIX, 4 (1957), pp. 138–42.

3 '... The Baconian method of induction... if consistently pursued, would have left science where it found it.' A. N. Whitehead, *Process and Reality* (New York; Macmillan, 1929), p. 7.

PAGE 103

1 W. H. Walsh, *An Introduction to Philosophy of History* (London; Hutchinsons' University Library, 1951), p. 103.

2 *Ibid.*

3 *Ibid.* p. 107.

4 *Ibid.*

PAGE 105

1 *Ibid.* p. 106.

2 'Common sense, as a system is laws, is delicately ramified, down to the nicest details of behaviour, as evidenced by the fact that we are so seldom surprised in our everyday doings and witnessings. Neither, of course, are infants often surprised: having *no* general notions, every experience is equivalently random, equally expected and unexpected, the infant being an unconscious master of the Principle of Insufficient Reason.' A. Danto, 'On Explanations in History', *Philosophy of Science*, XXIII, 1 (1956), p. 27.

3 Cf. H. P. Grice and P. F. Strawson, 'In Defense of a Dogma', *Philosophical Review*, LXV, 2 (1956), especially pp. 150 ff.

4 F. H. Bradley, 'The Presuppositions of Critical History', in *Collected Essays by F. H. Bradley*, vol. 1 (Oxford University Press, 1935).

PAGE 108

1 Walsh, *op. cit.* p. 108.

PAGE 112

1 Arthur Danto, 'On Historical Questioning', *Journal of Philosophy*, LI (1954), 89–99.

PAGE 113

1 'Since the history of any period embraces all the actualities involved, and since both documentation and research are partial, it follows that the total actuality is not factually knowable to any historian, however laborious, judicial, or faithful he may be in his procedures. History as it actually was...

is not known or knowable, no matter how zealously is pursued "the ideal of the effort for objective truth".' C. Beard, 'That Noble Dream', p. 324.

PAGE 115

1 And I am saying that nothing contrasts with this which is recognizably a piece of historical writing. For an analogous point, see the discussion in Christopher Blake, 'Can History be Objective?', *Mind*, LXIV (1955), 61–78, reprinted in P. Gardiner, *Theories of History*, pp. 329–43. Blake cautions us against using the word 'objective' to apply to accounts we cannot so much as imagine, not because an objective account is immeasurably difficult to produce, but because it is by no means clear what we would mean by 'objective account'. Blake writes, after remarking upon the indeterminacy of usage here, that 'we cannot say with any precision what an objective account of anything would be like' (p. 343). He reminds us that 'before we started to wonder, we did know how to use the word'.

PAGE 116

1 Benedetto Croce, *History—Its Theory and Practice*, tr. D. Ainslee (New York; Russell & Russell, 1960). See especially ch. 1, the bulk of which is reprinted in Gardiner, *op. cit.*

PAGE 117

1 W. H. Walsh, *Introduction to Philosophy of History*, p. 31. He writes, however, 'The point on which I want to insist is that, though it is possible to find these two levels of chronicle and history proper throughout written history —though it is possible to find elements of chronicle in the most sophisticated history, and of history proper in the most primitive chronicle—the historical ideal is always to get away from the stage of chronicle and attain that of history itself' (p. 33). I wish to insist, on the other hand, that there are not two kinds of things, portions of which may be found in every instance of historical narration. It is not even a distinction in kinds of activity such as, say, experimenting and theorizing are in physics.

PAGE 119

1 I mean that in the remarks to follow, I am not to be engaged in a strictly *ad hominem* argument against Walsh's views. I shall be taking Walsh's claims as general claims, and I shall be using them to make general points. Walsh has simply thought out with greater clarity and in greater detail certain notions which are very widely held indeed.

PAGE 120

1 Walsh, *op. cit.* p. 32.
2 *Ibid.* p. 33.

PAGE 122

1   Charles S. Peirce, *Collected Papers*, vol. v, para. 146. See particularly the discussion of abduction in N. R. Hanson, *Patterns of Discovery*, pp. 85 ff. For a comparable approach, based upon the falsificationist thesis of Karl Popper, and with specific application to history, see Joseph Agassi, *Towards an Historiography of Science*, printed as Beiheft 2 of *History and Theory* (1963).

PAGE 123

1   But we could hardly imagine the immense variety with which the Last Supper theme was in fact instantiated, only taking into consideration the series beginning with Castagno and ending with Veronese. This relationship between concept and instance is a critical one, and I shall discuss it at some length in connection with my analysis of historical explanation.

PAGE 124

1   'It is no light matter to write the history of painting in Greece', wrote Mary Hamilton Swindler in her important history of the subject, *Ancient Painting* (New Haven; Yale University Press, 1931), p. 109. To begin with, the great paintings are lost. But 'this is not to say that there is no painting left to use, nor that we can form no adequate idea of it' (p. 110). Again, few antique writings go back beyond the third century, and one of the main works, Pliny's, is marred by the fact 'that often he did not understand the authors from whom he drew'. But we apparently are in a position to show this. Finally, we have, for various reasons, the idea that painting was an essentially subordinate art in Greece, a fact which has inhibited us from correctly estimating the force of some of the data. Despite all these things, a narrative of Greek painting can be written.

2   Walsh, *op. cit.* p. 33.

PAGE 125

1   'In the year 1891, Manet and Seurat were already dead; Pissaro, Monet and Renoir were at their height of powers; Cézanne had opened yet another world. *Sunday at la Grande Jatte* and *Le Dejeuner dans le Bois*, *La Musique aux Tuileries*, *Les Dames dans un Jardin*, the ochre farms and tawny hills of Aix were there, on canvas, hung, looked at—to be seen by anybody who would learn to see. But were they seen ? . . . . For the age of the Impressionists was also still the age of decorum and pomposity, of mahogany and the basement kitchen, the overstuffed interior and the stucco villa; an age that venerated old, rich, malicious women and the clever banker; when places of public entertainment were large, pilastered and vulgar, and anyone who was neither a sportsman, poor, nor very young, sat down on a stiff-backed chair

three times a day eating an endless meal indoors.' Sybelle Bedford, *A Legacy*, III, 1.

2   The analogue in the case of memory is that memories do not decay with *time*, but rather as a function of the number of intervening experiencings increases. This is experimentally demonstrable.

PAGE 126

1   Ibn Khaldun, *An Arab Philosophy of History*, tr. and arranged by Charles Issawi (London; John Murray, 1950), pp. 31–2.

PAGE 128

1   See the discussion in P. F. Strawson, *Individuals: An Essay in Descriptive Metaphysics* (London; Methuen, 1959), pp. 20 ff.

PAGE 130

1   Walsh, *op. cit.* p. 32.

PAGE 131

1   Leopold von Ranke, *Preface to Histories of the Latin and German Nations from 1494–1514*. Tr. by the editor in Fritz Stern (ed.), *The Varieties of History*, pp. 55–60.

2   For example, by Pieter Geyl, *Debates with Historians* (New York; Meridian Books, 1958), ch. 1: 'Ranke *is* to be found in his work.'

3   Or that he meant to understand events in just the same way in which those who lived through them understood them. But then 'to understand Greece *wie es eigentlich gewesen* is not only impossible, but it is not even a valid idea of knowledge', writes J. H. Randall Jr. in *Nature and Historical Experience* (New York; Columbia University Press, 1958, p. 64). Randall gives no argument.

PAGE 140

1   W. H. Walsh, '"Plain" and "Significant" Narratives in History', *Journal of Philosophy*, LVIII (1958), 479–84. This is a reply to a paper of mine, 'Mere Chronicle and History Proper', *Journal of Philosophy*, L (1953), 173–82. This, in turn, is an earlier version of part of the present chapter.

PAGE 143

1   Irwin Lieb (ed.), *Charles S. Peirce's Letters to Lady Welby* (New Haven; Whitlock's, 1953), p. 9. Peirce says this in the midst of discussing his theory of Categories. This is complicated enough, but he is also giving *en passant* an account of the sorts of reasons which must have led Kant to the view that Time is a 'form of the internal sense alone'. From the context it is not clear

whether the sentence is asserted by Peirce or imputed by him to Kant, or whether he supposes that Kant implictly subscribed to it. It appears in a tangled, and inconsistent, discussion, but I am not examining Peirce's views as such, only using his statement as representative of a widely held view. Cf. '[People] . . . have very different pictures of the past and the future. The past is thought of as being "there", fixed, unalterable, indelibly recorded in the annals of time, whether we are able to decipher them or not. The future, on the other hand, is regarded as being not merely largely unknown but largely undecided. . . . Thus the future is thought to be open, whereas the past is closed.' A. J. Ayer, *The Problem of Knowledge* (London; Macmillan, 1956), p. 188.

PAGE 145

1   C. S. Peirce, *loc. cit.* p. 9.
2   I allude, of course, to P. F. Strawson, 'On Referring', *Mind* (1950), reprinted in A. Flew (ed.), *Essays in Conceptual Analysis* (London; Macmillan, 1956). I cannot accept Strawson's *general* thesis—see my 'A Note on Expressions of the Referring Sort', *Mind* (1958). So application of it to references to purported future occasions would have to be independently argued. The entire difficulty arises from the view that the truth or falsity of a sentence S is independent of the time at which S is uttered. Strawson must argue that sentences as such are never either true or false, only *statements* are; and whether these are true or false is very much a matter of the time at which they are asserted. But if we regard sentences without the appropriate temporal information as incomplete, we may then regard sentences when appropriately completed as true independently of the time of their utterance. But this solves none of the epistemological problems I am to be concerned with.
3   See the elementary discussion of this in A. Heyting, *Intuitionism: An Introduction* (Amsterdam; North Holland Publishing Co., 1956), pp. 1 ff. Heyting would justifiably rule out my 'extension' as 'metaphysical'.
4   C. D. Broad, *The Mind and its Place in Nature* (London; Kegan Paul, Trench, Trubner, 1925), p. 252.

PAGE 146

1   *Ibid.*

PAGE 147

1   Gilbert Ryle, *The Concept of Mind*, pp. 301–4 and *passim*.
2   For example, Bertrand Russell, *The Analysis of Matter* (London; Kegan Paul, Trench, Trubner, 1927), p. 294. '. . . No event lasts for more than a few seconds at most.' By 'event' Russell means a component of an object

having physical structure. On the other hand, 'Whether to call the Battle of Waterloo an event is a matter of words' (p. 293). But see M. Mandelbaum, *The Problem of Historical Knowledge* (New York; Liveright, 1938), p. 254 and *passim*. Mandelbaum regards the Reformation as an event. I shall later introduce the term 'temporal structure' for very large events.

PAGE 148

1 Galileo Galilei, *Dialogo sopra i due massimi sistemi del mondo*, in *Opere* (Florence; Edi. Naz., 1929–39), VII, 129.

2 I appreciate the fact that it is important for maps to be incomplete. 'For when our map becomes as large and in all other respects the same as the territory mapped—and indeed long before this stage is reached—the purposes of a map are no longer served. There is no such thing as an unabridged map; for abridgment is intrinsic to map making.' (Nelson Goodman, 'The Revision of Philosophy', in Sidney Hook (ed.), *American Philosophers at Work*: New York; Criterion Books, 1956, p. 84.) But of course *this* map is not an exact replica: there is as much difference between an event and its description as there is between Pittsburg and a dot. Moreover, the use to which my 'map' is to be put requires completeness.

3 I do not mean to suggest that these are the *only* problems regarding maps.

PAGE 150

1 Benedetto Croce, *History: Its Theory and Practice, passim*.

PAGE 151

1 In her book, *Intention* (Oxford; Basil Blackwell, 1957), G. E. M. Anscombe points out that there are many descriptions of an action, only under some of which is an action intentional. I think this a considerable insight, and I want to acknowledge that my own thoughts here were directly stimulated by Miss Anscombe's book.

PAGE 152

1 Alfred North Whitehead, *Adventures of Ideas* (New York; Macmillan, 1933), p. 246.

PAGE 153

1 See Mandelbaum, *op. cit.* chs. I and IV.

2 Max Black, 'Why Cannot an Effect Precede its Cause?', *Analysis*, XVI (1956), 49–58.

PAGE 155

1 For familiar reasons. By definition, p states a necessary condition for q if $\sim p \supset \sim q$. But this is equivalent to $q \supset p$. And this exactly represents the

claim that q is a sufficient condition for p. In brief, whenever p is a necessary condition for q, q is a sufficient condition for p, and conversely.

2 Though of course the so-called mechanical state of a physical system s determines every other state of s for every value of t—including all temporally earlier states of s.

PAGE 158

1 N. R. Hanson would argue that we don't see the same thing they saw, that even, say, a contemporary historian of science and his wife who is totally uninterested in the history of science would not, parity of retinal images notwithstanding, see the same thing when both view the house. See his *Patterns of Discovery* especially ch. 1.

PAGE 164

1 Of course, if B_i is admitted into the range $B_1 \ldots B_n$ marked out by 'is R-ing', this is doubtless because of some strong evidence that B_i in general leads to R, or that failure to B_i leads to a failure of R. Indeed, if one may speculate on the history of language, it may very well be that project-words get applied to various actions in this way. But once the convention is part of common usage, ascription of B_i does not entail the prediction that R.

PAGE 167

1 This is argued in detail in my 'Mere Chronicle and History Proper', *Journal of Philosophy*, L (1953).

PAGE 168

1 Cecil Woodham-Smith, *The Reason Why* (New York; McGraw-Hill, 1954), p. 167.

2 *Ibid.* It is only necessary to pick a history book at random to find examples of this manner of speaking. Thus: 'At the very moment when it seemed that the Papacy should have concentrated all its forces to resist its enemies, it flung itself into the crisis which is known as the Great Schism, and which for forty years was to rend Western Christiandom in twain.' (Henri Pirenne, *History of Europe*: New York; Anchor Books, 1956, II, 122.) 'A disagreeable incident occurred as Erasmus was leaving English soil in January 1500. . . . Yet this mishap had its great advantage for the world, and for Erasmus too, after all. To it the world owes the *Adagia*; and he the fame, which began with this work.' (J. Huizinga, *Erasmus and the Age of Reformation*: New York; Harper Torchbooks, 1958), pp. 34–5.) 'And yet this business, so distasteful in itself, was of supreme importance in world history. This Church, with its collateral sects grown rigid and cut off from all development, was for another millenium and a half to hold nationalities together against the pressure of the barbarians, even to take the place of

nationalities, for it was stronger than state or culture, and therefore survived them both. In it alone there persisted the essence of Byzantism.' (Jacob Burckhardt, *The Age of Constantine the Great*: New York; Anchor Books, 1954, p. 302.) '[Oresme's] work was a step towards the invention of analytical geometry and towards the introduction into geometry of the idea of motion which Greak geometry had lacked.' (A. C. Crombie, *Augustine to Galileo: The History of Science: A.D. 400–1650*: Cambridge, Mass.; Harvard University Press, 1953, p. 261.) This last example (and they could be multiplied endlessly) is quoted in an important paper by Joseph T. Clark, 'The Philosophy of Science and the History of Science', in Marshall Clagett (ed.), *Critical Problems in the History of Science* (Madison, Wisconsin; The University of Wisconsin Press, 1959), p. 127. All my examples are instances of what Father Clark terms *die von unten bis oben geistesgeschichtliche Methode*, a method particularly susceptible to what he terms 'precursitus' (*loc. cit.* p. 103, and note 2, p. 138). Precursitus (if it be a lapse), and the whole *Methode* characterized by Father Clark, are due to narrative description, a mode of description which goes *von später bis früher*.

3 'Men who never think independently have nevertheless the acuteness to discover everything, after it has once been shown them, in what was said long since, though no one ever saw it there before.' Immanuel Kant, *Prolegomena To any Future Metaphysic*, para. 3.

4 Henri Bergson, *La Pensée et le Mouvant* (Paris; Felix Alcan, 1934), p. 23. This passage is cited by Mandelbaum, *op. cit.* p. 29. I am indebted to Professor Mandelbaum for drawing my attention especially to Bergson's discussion.

PAGE 169

1 Perhaps this sentence, though grammatically one, breaks up into a conjunction which contains a sentence in the future tense as one conjunct. Thus, it asserts: (*a*) Aristarchus did such and such at *t-1*; (*b*) Copernicus will do such and such at *t-2*; (*c*) *t-1* is earlier than *t-2*; (*d*) so-and-so resembles such and such. But (*b*) shifts tense after 1543, and this confirms the point I make below.

2 This is perhaps questionable. Consider the case of lying. A man intends *S* to be a lie, but unbeknownst to himself he utters a true sentence. Shall we say he lied anyway, the intention to lie being enough to make of *S* a lie? Or shall we say that he *tried* or *meant* to lie, and failed? I would say the latter. And similarly I would say the man tried and failed to predict. But this may be simply legislation on my part.

3 Even this wants amplification. Suppose *E* never happens, so that I cannot stand in any temporal relation to *E*: I suggest that there must be some implicit time limitation, e.g. at *t-1 E* is predicted to occur at *t-2*, so the *full*

20   305

prediction is '*E*-at-*t*-2'. If *E* fails to occur at *t*-2, the prediction will be false. But obviously we cannot always make such specifications. I may predict that I will die, but save for special contexts, the date is hidden from me.

PAGE 173

1 See below.
2 Ludwig Wittgenstein, *Philosophical Investigations*, p. 223e.

PAGE 183

1 Some historians, of course, appreciate the fact that their temporal distance from an event constitutes an advantage rather than an occupational liability. Thus: 'Il ne faut pas projeter indûment les développements ultérieurs sur la situation précédente, rendre par example Platon "respons-able" du skepticisme de la Nouvelle Académie, ni St-Augustin de Jan-sénius. Mais l'effort même qui me conduit à établir que le Jansénisme est un développement bâtard de l'Augustinisme m'aide puissamment à mieux comprendre ce dernier.' H.-I. Marrou, *De la Connaissance Historique* (Paris; Editions de Seuil, 1959), pp. 46–7.

PAGE 184

1 For plainly it is not a sufficient condition for an action *a* to be free that *a* be intentional.

PAGE 187

1 Gilbert Ryle, *Dilemmas* (Cambridge University Press, 1956), p. 21.
2 Aristotle's opponent is cleverer than Ryle's. 'It makes no difference whether people have or have not actually made the contradictory statements. For it is manifest that the circumstances are not influenced by the fact of an affirmation or denial on the part of anyone. For events will not take place or fail to take place because it was stated that they would or would not take place, nor is this any more the case if the prediction dates back ten thousand years or any other space of time. Wherefore, if through all time the nature of things was so constituted that a prediction about an event was true, then through all time it was necessary that that prediction should find fulfilment.' Aristotle, *On Interpretation*, 18*b*–19*a*, tr. E. M. Edgehill.
3 Denis Diderot, *Jacques le Fataliste* (Paris; Bibliothèque Mondiale, n.d.), p. 19.

PAGE 188

1 *Ibid.* p. 17.

PAGE 189

1 'Yet this view leads to an impossible conclusion; for we see that both deliberation and action are causative with regard to the future, and that, to

speak more generally, in those things which are not continuously actual there is a potentiality in either direction.' Aristotle, *op. cit.* 19*a*.

2   This move is suggested, as is much of my general presentation of the argument thus far, by the interesting discussion in Colin Strang, 'Aristotle and the Sea Battle', *Mind*, LXIX (1960), p. 463. I must emphasize, however, that my argument deviates considerably from Strang's hereafter, and that I attempt rather to reconstruct Aristotle than to defeat his opponent. That is to say, if the opponent is wrong, the problem remains of giving a positive account, which is what I seek to do.

PAGE 190

1   Richard Taylor, 'The Problem of Future Contingencies', *Philosophical Review*, LXVI (1957), p. 3. Taylor gives, in n. 2, a good bibliography of recent discussions on this subject; and part, but only part, of the present analysis is, I think, compatible with Taylor's own excellent account.

2   *On Interpretation*, 18*b*.

PAGE 191

1   *Ibid.*

PAGE 192

1   That is, though neither '*s* will be *F*' nor '*s* will not be *F*' is true or false, the *disjunction* of the two is true. Actually, as I formulate it, it need not be true. Supposing there will be a sea-battle tomorrow, we may say that it will be fatal to the one side, or it will not be. But it need not be fatal nor otherwise if there in fact turns out to be no sea battle. To be sure, 'There will be a sea battle tomorrow or there won't be a sea-battle tomorrow' is perhaps necessarily true (providing they both refer to the *same* tomorrow and that that tomorrow ever comes).

PAGE 193

1   That is, if *S* did take place.

2   That is, if *S* did take place, *S* was either one of *F* or not-*F*.

PAGE 196

1   See the superb discussion in A. N. Prior, *Formal Logic* (Oxford; Clarendon Press, 1955), part III, ch. II, sect. 2.

PAGE 197

1   Richard Taylor, *loc. cit.* p. 26.

PAGE 203

1   C. G. Hempel, 'The Function of General Laws in History', *Journal of Philosophy*, XXXIX (1942). Reprinted frequently, but I shall make references

to P. Gardiner (ed.), *Theories of History*. It is fair to say that almost everything since published on the topic has been structured by Hempel's original formulation, whether writers agree with him or not. For a bibliography of post-war literature, see John C. Rule, *Bibliography of Works in the Philosophy of History, 1945–1957*, published as Beiheft 1 of *History and Theory* (1961).

PAGE 204

1 It might seem as though one could hardly take (3) to be common ground and then go ahead and deny (1): for how is one to ascertain that in fact historians' explanations contain no general laws when, if (1) is false, there are no such explanations? (If (3) is true, (1) must be, and so (2) must be false.) Nevertheless, when people assert (3) and deny (1), they have in mind sufficient qualifications to permit this, as will come out in the following discussion.

PAGE 205

1 I speak of Historical Idealists, rather than Historicists, for that word is either too vaguely bestowed or else is too precisely associated with the use given it by Professor Karl Popper, whose views I should wish to quarantine as much as possible, they being rather peculiarly his own.

PAGE 209

1 C. G. Hempel and P. Oppenheim, 'The Logic of Explanation', *Philosophy of Science*, XV (1948). Reprinted in H. Feigl and M. Brodbeck, *Readings in Philosophy of Science* (New York; Appleton, Century & Crofts, 1953).

PAGE 210

1 Hempel, 'The Function of General Laws in History', *loc. cit.* p. 351.
2 See Karl Popper, *The Poverty of Historicism* (Boston; Beacon Press, 1957), p. 144 and *passim*.

PAGE 213

1 Michael Scriven, 'Truisms as Grounds for Historical Explanations', in P. Gardiner (ed.), *Theories of History*; and 'Explanations, Predictions, and Laws', in H. Feigl and G. Maxwell (eds.), *Scientific Explanation, Space, and Time, Minnesota Studies in the Philosophy of Science*, vol. III (Minneapolis; University of Minnesota Press, 1962), pp. 170 ff.
2 Karl Popper, *The Open Society and its Enemies* (Princeton; Princeton University Press, 1950), pp. 448 ff. 'If we explain, for example, the first division of Poland in 1772 by pointing out that it could not possibly resist the combined power of Russia, Prussia, and Austria, then we are tacitly using some trivial universal law such as "If of two armies which are about equally well armed and led, one has a tremendous superiority in men, then the

other never wins." . . . Such a law might be described as a law of the sociology of military power; but it is too trivial ever to raise a serious problem for the students of sociology, or to arouse their attention.'

3   Ernest Nagel, *The Structure of Science* (New York; Harcourt Brace & World, 1961), *passim*.

4   William Dray, *Laws and Explanations in History* (Oxford; Oxford University Press, 1957), p. 57.

PAGE 214

1   Of course, that an event may be covered by a law, and that a description of an event may be deduced from explanatory premisses containing a general law, may be regarded as distinct theses. They are so treated by Alan Donagan in 'The Popper-Hempel Theory of Historical Explanation', *History and Theory*, IV, 1 (1964). Donagan defends the deductivist account, but rejects the covering law account.

2   Dray, *op. cit.* pp. 66 ff.

PAGE 215

1   See, e.g. William Dray, '"Explaining What" in History', in P. Gardiner (ed.), *Theories of History*.

2   For a rather similar point, see J. Passmore, 'Explanation in Everyday Life, in Science, and in History', *History and Theory*, II (1962), 2.

PAGE 217

1   For if the explanans *Es* could be true simultaneously with the explanandum *Em* being *false*, in what sense would we have explained *Em* with reference to *Es*? Not, of course, that this is enough to establish that an explanation requires a deductive operation, but only that if this is not satisfied, how are we to speak of any kind of explanation at all?

2   Hempel and Oppenheim, 'The Logic of Explanation', *loc. cit.* III, para. 6.

PAGE 224

1   Most notably by Israel Scheffler, 'Explanation, Prediction, and Abstraction', *British Journal for the Philosophy of Science*, VII (1957), 28. Reprinted in A. Danto and S. Morgenbesser (eds.), *Philosophy of Science*, pp. 274 ff.

PAGE 225

1   See Alan Donagan, *loc. cit.* sect. 7. Donagan, I believe, is not an historicist.

PAGE 228

1   Alan Donagan, 'Explanation in History', *Mind*, LXVI (1957). Reprinted in P. Gardiner (ed.), *Theories of History*. (See p. 430.)

PAGE 229

1   In this regard it is not easy to see how Donagan can, consistently with his assumption that historians explain, go on to affirm the deduction requirement, and at the same time reject the covering law requirement.

PAGE 234

1   Patrick Gardiner, *The Nature of Historical Explanation* (Oxford; Oxford University Press, 1952).
2   Ernest Nagel, *The Structure of Science*, pp. 564 ff.

PAGE 235

1   *Ibid.* p. 565.

PAGE 237

1   Bertrand Russell, 'On the Notion of Cause, with Applications to the Free-will Problem', in H. Feigl and M. Brodbeck (eds.), *Readings in Philosophy of Science*.
2   C. G. Hempel, 'The Function of General Laws in History', *loc. cit.* p. 346.

PAGE 239

1   Where, in my own idiom, we proceed from *conceptual* to *documentary* evidence.

PAGE 240

1   C. V. Wedgwood, *The Thirty Years War* (Penguin Books, 1957), p. 167.

PAGE 241

1   G. M. Trevelyn, *England Under the Stuarts* (New York; 1906). Cited in Nagel, *op. cit.* p. 564–5.

PAGE 242

1   See A. Danto, 'On Historical Questioning', *loc. cit.*

PAGE 246

1   Thus you cannot say that Louis XIV died unpopular because he ate poisoned lobster: for *that* only explains why he *died*. Indeed, it would not even enter into an explanation of why he died unpopular.

PAGE 249

1   W. B. Gallie, 'Explanations in History and the Genetic Sciences', reprinted in Gardiner (ed.), *Theories of History*, pp. 386 ff.

PAGE 259

1   J. W. N. Watkins, 'Historical Explanation in the Social Sciences', *British Journal for the Philosophy of Science* (1957). Reprinted in P. Gardiner (ed.), *Theories of History*. Page references are to the Gardiner volume. The passage cited is there on p. 505.

2   H.-I. Marrou, *De la Connaissance Historique* (Paris; Editions du Seuil, 1959), p. 177. Strictly speaking the cited sentence is plainly false: experience furnishes examples of masses of organisms other than human individuals. It indeed furnishes us with examples of organisms which have other organisms as parts of themselves. It is a moot point whether it furnishes examples of organisms which have human individuals as parts of themselves. But surely it is only this point that Marrou is concerned with.

PAGE 260

1   *Ibid.*
2   Watkins, *loc. cit.* p. 505.

PAGE 261

1   C. V. Wedgwood, *The Thirty Years War* (Pelican Books, 1957), p. 339.

PAGE 262

1   *Ibid.* pp. 339–40.

PAGE 263

1   *Ibid.*
2   *Ibid.*

PAGE 264

1   Watkins, *loc. cit.* p. 511.

PAGE 265

1   H.-I. Marrou, *op. cit.* p. 177. Cf. pp. 163 ff.
2   Watkins, *loc. cit.* p. 505.
3   The literature on Methodological Individualism is substantial, but much of it fails to make the distinctions I feel are crucial. For background, see Watkins's earlier paper, 'Ideal Types and Historical Explanation', *British Journal for the Philosophy of Science* (1952), reprinted in H. Feigl and M. Brodbeck, *Readings in the Philosophy of Science* (New York, 1953). Criticisms may be found in M. Mandelbaum, 'Societal Facts', *British Journal for the Philosophy of Science* (1955); L. Goldstein, 'The Inadequacy of the Principle of Methodological Individualism', *Journal of Philosophy* (1956); E. Gellner, 'Holism versus Individualism in History and Sociology,' *Proceedings of the Aristotelian Society* (1956). The papers of Mandelbaum (which I shall discuss

on p. 270 ff.) and of Gellner are both reprinted in Gardiner, *op. cit.* which also contains a brief note—'Reply to Mr. Watkins'—by Gellner. In addition see the earlier discussions. See F. A. Hayak's two books, *Individualism and the Economic Order* and *The Counter Revolution of Science*, as well as Karl Popper's *The Open Society and its Enemies* and *The Poverty of Historicism*.

PAGE 270

1 Hence Professor Popper is perfectly justified when, in ch. 14 of *The Open Society and its Enemies*, he condemns Psychologism and recommends Methodological Individualism, and Mr Gellner is in logical error when he seeks to rebut Methodological Individualism through arguing against Psychologism, feeling that the two cannot be distinguished. See 'Holism versus Individualism in History and the Social Sciences', in Gardiner, *op. cit.* 501 and n. 9.
2 Page references will be to the Gardiner volume.
3 *Ibid.* p. 479. I do not believe that the adjective 'specific' is to be taken as a rejection of reference to anonymous individuals.
4 *Ibid.* p. 478.
5 *Ibid.* p. 483.

PAGE 271

1 *Ibid.*
2 *Ibid.* p. 486.
3 *Ibid.* p. 488. A philosophical puritan would strenuously object to the notion that facts are able to interact, or that facts can possibly come into conflict. *Things* can interact, things and propositions can come into conflict. But this relaxed use of 'fact' does not intrinsically damage Mandelbaum's argument, and can readily be reconstructed.
4 *Ibid.* p. 481.

PAGE 272

1 *Ibid.* p. 482. Should the 'theoretical possibility', though 'practical impossibility', of such a translation be demonstrated this would 'be significant from the point of view of a general ontology, but would not affect my argument regarding the autonomy of the societal sciences'.

PAGE 276

1 Watkins, 'Historical Explanation in the Social Sciences', p. 509.
2 *Ibid.*

PAGE 279

1 See in particular ch. 11 of his *The Structure of Science: Problems in the Logic of Scientific Explanation* (New York; Harcourt, Brace & World, 1961).

PAGE 283

1   'What I regard as the real issue is something like this: social scientists can
     be roughly and crudely divided into two main groups: those who regard
     social processes as proceeding, so to speak, under their own steam, accord-
     ing to their own nature and laws, and dragging the people involved along
     with them; and those who regard social processes as the complicated out-
     come of the behavior of human beings.' J. W. N. Watkins, personal
     letter, dated 11 January 1962.

INDEX

Achievements, 8, 84–5, 147
Agassi, Joseph, 300
Analytical philosophy of history, 1, 15, 16, 93
Anscombe, G. E. M., 303
Aristotle, 11
  on future contingencies, 189–200, 306, 307
Atomic narratives, 251–3
Ayer, A. J., 30, 34, 293–6, 302
  on verification, 45
  on verifiability in principle, 46–53, 58–61
  on analysis of tensed sentences, 47–61

Bacon, Francis, 101, 298
Beard, Charles, 33, 88–102, 104, 111, 113–15, 296, 297, 299
Becker, Carl, 33
Bedford, Sybelle, 301
Beginnings, 245–56
Bergson, Henri, 168, 305
Bernard, Claude, 96, 297
Black, Max, 303
Blake, Christopher
Bradley, F. H., 105–6, 298
Broad, C. D., 145, 302
Burkhardt, Jakob, 15, 288, 305
Butler, Ronald, 89, 295, 296

Causes
  do not succeed their effects, 153–5
  of changes, 245–56
Causality, vindication of Hume's theory of, 242–5
Changes
  as explananda in narration, 233–56
  in social individuals, 261, 262, 263, 264
  in the past, possibility of, 153–5, 181

Chronicle, 115–42
Clark, Father Joseph, 305
Collingwood, R. G., 205
Continuous Series Model, 214
Correspondence Theory of Truth, 186, 199
Covering Law Model, 214–15
Croce, Benedetto, 33, 116, 205, 299, 303
Crombie, A. C., 305

Danto, Arthur C., 298, 310
Deduction assumption, 206–8
De Interpretatione, 189–200
Descriptions, as presupposing general laws, 218–23, 227
Determinism, 182–6
  Logical determinism, 186–200
Dewey, John, 30, 37, 81, 292
Dialectical pattern, as resembling narrative structure, 237
Diderot, Denis, 187–9, 197, 306
Dilthey, Wilhelm, 205
Disagreements of principle, 109–10
Donogan, Allen, 228, 304
Dray, William, 213–15, 216, 222, 227, 234, 236, 309
Duhem, Pierre, 97, 101, 297, 298

Eliot, T. S., 288
Engels, Friedrich, 285–6
  See also Marx, Karl
Event, looseness in the concept of, 2, 147
Evidence, 37, 40, 89–94
  conceptual and documentary evidence, 122, 125–9, 139, 226
  and prediction, 171–81
Explanata, 221–3, 225, 228, 230
Explanations, historical, Chs. IX, X, passim

Index

Explanation sketches, 209, 210–11, 238–9, 272

Fatalism, 187
Free-will, in history, 184–6
Freud, Sigmund, 33, 291
Full descriptions of events, 148–53, 155, 219–20
Future-referring terms, 71, 75–6, 164

Gale, Richard, 292
Galileo, Galilei, 303
Gallie, W. D., 310
Gardiner, Patrick, 234, 236, 286, 310
Geisteswissenschaften, 205
Gellner, Ernest, 311, 312
General Laws, 208, 209, 212
  and descriptions, 218–23, 224–32
  and social systems, 267, 275
  in narrative explanation, 237–41, 252–266
  in reduction, 277–84
Geyl, Pieter, 301
Goldstein, Leon, 311
Goodman, Nelson, 303
Grundy, G. B., 289

Hanson, N. R., 297, 300, 304
Hayak, F. A., 312
Hegel, G., 14, 237, 268, 285, 287
Hempel, C. G., 203, 206–10, 211, 213, 215–16, 218–19, 224–5, 237, 245, 295, 307, 308, 310
Herbst, P., 295
Heyting, A., 302
Historical foreknowledge, 182, 196–7, 200
Historical idealists, 205, 308
Historical laws, 254–6
Historical meaning, 7–15, 17
  See also Significance
Historical novels, 63
Historical questions, 112, 137–8
Historical sentences, 258–9

Historical wholes, 8
  See also Temporal wholes
Historical, minimal characterization of, 25
Hitler, Adolf, 287
Holland, R. F., 292, 293
Hospers, John, 293
Huizinga, J., 301
Hume, David, 23–4, 40, 76–7, 103–6, 242–5, 290, 292, 294, 295

I-predicates, 272–5
Ibn Khaldun, 126–7, 301
Ideal Chronicle, 2, 149–82
Illusions of explanation, 224, 227, 229, 232
Imagination, historical, 121
Induction, 20–4
Instantaneous scepticism, 84–5
Instrumentalism, historical, 79–85, 295–6
Intentions, 182–6, 263, 284

Jacques le Fataliste, 187–9

Kant, Immanuel, 3, 60, 257, 286, 301, 305
Kepler, Johannes, 3, 4, 5

Lewis, C. I., on statements about the past, 30, 34–44, 45, 60, 99, 291, 292
Liddell and Scott, 289
Leibniz, G., 263
Logical determinism, 186–200
Löwith, Karl, 7, 9, 286

MacDonald, Margaret, 32, 291
Macro- and microscopic descriptions, analogues of in social science, 279–83
Mandelbaum, Maurice, 270–5, 303, 305, 311, 312
'Marks of pastness', 39–40, 44, 66, 72
Marrou, H-I, 259, 260, 264, 265, 282, 306, 311
Marxism, 2, 268–70, 285, 286
Marx, Karl, 3, 9, 269

Index

Masi, Evelyn, 291
Memory, empiricist theory of, 33, 39–40, 63, 66–7
Methodological socialism, 268–84
Methodological individualism, Ch. XI, *passim*
Molecular narratives, 252–5

Nagel, Ernest, 98, 213, 234–5, 297, 309, 310
Narratives, as historical theories, 121, 137
   'plain' and 'significant', 116–42
   plausible, 123, 127–9
   role of in historical explanation, 251–6
   unity of, 248–56
Narrative explanation, model of, 236
Narrative organization of events, 142
Narrative sentences, Ch. VIII, *passim*, 182–6, 188, 194–6, 197, 202, 234–5
Naturwissenschaften, 205–6
Newton, Isaac, 3, 4
Nietzsche, Friedrich, 33, 291

Omniscience, 196–7
Oppenheim, Paul, and Hempel, C. G., 308, 309
Other possibilities, in the class and membership senses, 228–31, 240

Passmore, John, 309
Past, difficulties in the phenomenalist rendering of, 50–2
   questionable importance of concept of, 66–9
Past contingencies, 196
Past-referring terms, 71–5, 78, 83–4, 86, 92, 93
Pears, David, 296
Peirce, Charles, 30, 243–5, 290, 295, 300, 301–2
Phenomenalism, 48–53, 57, 60, 99
Pirenne, Henri, 304
Plato, 22

Popper, Karl, 210, 213, 268, 286, 287, 300, 308, 312
Precognition, 70, 174–5
Prediction, 9, 12, 169, 185–6
   and explanation, 225–32
   and prophecy, 286–7
Presley, C. F., 289
Price, H. H., 66, 294
Prior, A. N., 294, 307
Projects, 161–70, 183
Prophecy, 9, 12, 18, 175, 255, 286–7
Proust, Marcel, 93–4
Psychologism, 270, 276

Randall, John Herman, Jr., 301
Ranke, Leopold von, 130–1, 133, 139, 301
Reduction, 261, 266, 278, 279–83
Reference, 59, 60, 69, 81, 152
Regret, 10
Relativism, historical, Ch. VI, *passim*
Russell, Bertrand, 30, 31, 40, 78, 192, 237, 266, 290–1, 292, 293, 294, 302, 310
Ryle, Gilbert, 84, 187, 296, 302, 306

Scepticism regarding the past, 28, 30, 31, 59, 63–5, 76–87
S-predicates, 272–5
Scheffler, Israel, 291, 295, 307
Scriven, Michael, 213, 215, 216, 217, 223, 308
Significance, 159, 167, 202, 263, 284
   some senses of, 132–9,
   See also Historical meaning
Social individuals, 258–70, 283–4
Societal facts, 270–5
Spengler, Oswald, 288
Stories, 11, 12, 201–3, 233
   See also Narratives
Strang, Colin, 307
Strawson, P. F., 30, 291, 298, 301, 302
Substantive philosophy of history, Ch. I, *passim*, 17, 18–19, 20, 26, 76, 255, 257
Swindler, Mary Hamilton, 300

Index

Taylor, Richard, 23, 190, 197, 289, 296, 307

Temporal provincialism, 126, 142

Temporally-neutral terms, 71–5, 78, 83–4, 90–3

Temporal wholes, 166–70, 183, 235, 248, 255–6

Tenses, 51, 52, 53–8, 73–4, 191

Tenseless idiom, 51, 52, 56–8, 59, 197–200

Theories, 80, 176–81, 195
    descriptive and explanatory, 2–4
    historians' use of criticised by Beard, 99–102

Theoretical terms, 79

Thermodynamics, and reduction, 279–82

Thucydides, 19–25, 288, 289

Time-falsehood and time-truth, 193–8

Toynbee, Arnold, 259

Trevelyan, G., 241, 310

Verifiability criterion of meaning, 29

Verifiability in principle, 46–9, 59–61, 63

Verificationism, 45, 49, 52, 264

Verstehen, 169, 206

Vico, Giambattista, 92

Walsh, W. H., on relativism, 102–10
    a history and chronicle, 116–42, 297, 298, 299, 300, 301

Waters, Bruce, 293

Watkins, J. W., 259, 260, 264, 265, 266, 268, 275–6, 280, 311, 312, 313

Wedgewood, C. V., 241, 262, 264, 284, 310, 311

Whitehead, Alfred, 152, 298, 303

Williams, Donald, 287

Witnessing events, 61–2, 151, 155–9, 164, 170–5, 183

Wittgenstein, Ludwig, 48, 60, 173, 227, 286, 297, 306

Woodham-Smith, Cecil, 304

Woolf, Virginia, 152

Yeats, William Butler, 151